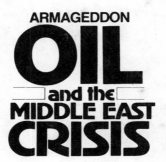

ARMAGEDDON

OIL
and the
MIDDLE EAST
CRISIS

THE BOOKS OF DR. WALVOORD . . .

Armageddon, Oil and the Middle East Crisis

The Blessed Hope and the Tribulation

The Church in Prophecy

The Holy Spirit

Israel in Prophecy

The Millennial Kingdom

The Nations in Prophecy

The Rapture Question

The Return of the Lord

The Thessalonian Epistles

ARMAGEDDON
OIL
and the
MIDDLE EAST
CRISIS

John F. Walvoord and **John E. Walvoord**

ZONDERVAN
PUBLISHING HOUSE
OF THE ZONDERVAN CORPORATION | GRAND RAPIDS. MICHIGAN 49506

Armageddon, Oil and the Middle East Crisis

© 1974, 1976 by The Zondervan Corporation,
Grand Rapids, Michigan

Sixteenth printing   February 1980

Library of Congress Catalog Card Number 74-4946

ISBN 0-310-34202-3

Quotations from the Bible are from *The New American Standard Bible*
© 1971 by the Lockman Foundation, published by Creation House,
Inc., and from *The New International Version* © 1973 by the New York
Bible Society, published by The Zondervan Corporation.

Printed in the United States of America

Contents

PREFACE   7

1. QUESTIONS ABOUT THE FUTURE   11
   The Incredible Twentieth Century/Rapid
   Change/The Impending Crisis/The Voice of
   the Prophets/The "Times" Are Running Out/
   Search for Answers

2. ARMAGEDDON CALENDAR   17
   No Vague and General Predictions/Seven Years
   Plus . . ./Will Israel Survive?/Oil and the New
   Power Grab/The Modern Dark Ages/Armaged-
   don Countdown/From World War III to
   Armageddon

3. THE ISRAELI-ARAB CONFLICT   25
   The Slender Thread of Hope/Freedom—Land—
   Prosperity/Arab Goal: Destroy Israel/War of
   1956: Arabs in Retreat/Russia Turns Against
   Israel/No Help From the United Nations/The
   Six Day War/Israel's Miraculous Victory/Rus-
   sia Attempts to Call a Cease-Fire/Final Days of
   War in 1967/Nasser's Myth of American and
   British Intervention/Results of the War of 1967
   /Russia Restocks Egypt and Syria/Egypt Breaks
   With Russia/Russia Repairs Break in Relations/
   The Arab Oil Blackmail/War Preparation in

1973/Early Arab Victories/World Reaction/
Mounting Casualties of the War/The Tide
Turns in Israel's Favor/Russia Presses for a
Cease-Fire/New Realities in the Middle East/
The Miracle of Israel's Survival/A Note of
Warning/The Bible Proved True Again

4. THE ARAB OIL BLACKMAIL   41

New Arab Unity/The World Energy Crisis/
The Vital Role of Middle East Oil/Oil at What
Cost?/United States Oil Potential—Going It
Alone/United States Problems—Money, Time,
and Ecology/The Unhappy Jew/From Black-
mail to Wealth and Power/End-Time Prophecy
in the Making/The United States and Russia in
Decline/The Prophetic Destiny of the Middle
East/The Coming Crisis — How Far Away?

5. PALESTINE: THE LAND OF PROMISE
      AND TRAVAIL   59

The Horizontal View of Prophecy/Prophetic
Questions About the Promised Land/Abram,
the Man of Destiny/The Promise of the Land/
Delayed Possession of the Land/The First De-
parture From the Land/Israel in Egypt/The
First Return/Immediate Occupation Condi-
tioned on Obedience/The Second Departure:
The Captivities/The Promise of the Second
Return/The Prayer of Daniel: The Second Re-
turn/The Temple Rebuilt/The Temple in the
First Century A.D./Destruction of the Temple/
The Third Departure/Prophetic Hope for a Last
Return/Scattered Israel's Persecutions/Begin-
nings of the Third Return/The Balfour Declara-
tion, 1917/The State of Israel Established in
1948/Israeli-Arab War, 1948-1949/Have the
Prophecies Been Fulfilled?/No Easy Road to
Glory/A New Kingdom Completes the Return/
Setting the Stage

6. THE CITY OF THE PROPHETS   77

Hallowed Memories/The City of Destiny/The
Ancient History of Jerusalem/The Center of
Israel's Life and Hope/The Messiah to Come
to Jerusalem/The Unwelcome King/Jesus De-
nounces the Religious Leaders of His People/
Jesus Predicts the Destruction of Jerusalem/The
Disciples' Questions/Jesus' Answers to the Disciples' Questions/The Present Age As Described
by Jesus/The Accuracy of Jesus' Predictions

7. WATCH JERUSALEM   89

What Is the Future of Jerusalem?/The War of
1967/Daniel's Prophetic Outline of World His-
tory/The Gap in Daniel's Prophetic Vision/
Confirmation of Daniel's Prophecy by the Apostle John/The Coming Desecration of the Holy
City/The Prophetic Pulse of Jerusalem/Chart of
Prophetic Events in History Beginning With
the Babylonian Captivity

8. THE RISING TIDE OF WORLD RELIGION   101

The Role of Religion in the End Time/Present
Religious Tensions in the Middle East/The
Church's Claim to Jerusalem/Is a United World
Religious Effort Possible?/Beginnings of the
Super-Church/The Prophetic Warning Against
False Prophets/The Rise of the Occult/What
Happens Next?/The Emergence of the Super-
Church/Communism As the Key to the Final
World Religion/Never Before

9. THE COMING MIDDLE EAST PEACE   113

The World Clamor for Peace/Peace and the
Final Countdown/The Middle East Market/
Israel Ready to Negotiate/A New World Leader
to Emerge/Conditions of the Settlement/Made
to Be Broken/ Waiting for the Prince of Peace

10. **THE THREAT OF RUSSIAN INTERVENTION**   121

The Fear of Russian Invasion/The Declining Power of the Ruble/The Russian Diplomatic Gamble/Russia's Future Exclusion From Middle East Affairs/Russia's Coming Invasion of Israel/Russia's Downfall/Is the Russian Invasion Near?/Russia's Loss—The Antichrist's Gain

11. **THE COMING WORLD DICTATOR**   133

The Dream of Conquerors/A Shrinking World/ The Advent of the Bomb/World Government— Man's Last Hope/Daniel's Prophecy of World Empires/The Rise and Decline of the Roman Empire/The Future Revival of the Fourth Kingdom/The Coming Dictator's First Move/Understanding Daniel's Seventy Weeks/The Counterfeit Prince of Peace/Forty-Two Months of Horror/Tools of World Domination Exist Today/ World Dictatorship Possible Today

12. **THE DAY OF WORLD CATASTROPHE**   147

Impending Nuclear Disaster/Increasing Pollution/Overpopulation and Starvation/The Threat of Lawlessness/The Coming Time of Jacob's Trouble/Daniel's View of the Time of Trouble/ The World's Time of Great Tribulation/Christ's Prophecy of the Great Tribulation/The Human Race Almost Destroyed/The Miraculous Survival of God's People/Details in the Book of Revelation/Seven Seals of Horror/Seven Trumpets of Disaster/Seven Last Bowls of Wrath

13. **ARMAGEDDON: THE WORLD'S DEATH STRUGGLE**   161

Proclamation of a New World Government/ The Rush to Judgment/The Outbreak of World War III/Revolt From the South and North/ Red China Makes Her Move/Satan's Battle Plan/The Satanic Trinity/The Battle of the

Great Day of God/Armageddon/To Curse God and Die

14. CHRIST'S SECOND COMING TO EARTH   169

From Manger to Power and Glory/As Lightning From East to West/The Whole World Awestruck/The Rider on the White Horse/World War III Ended/A Personal Return/A Visible and Glorious Return/A Return With the Heavenly Host/A Return Which Shakes the Earth/A Return to Judge the Nations/A Return to Rule From Zion/Living Jews Regathered and Judged/Other Survivors Face Judgment/From Disaster to Utopia/The Question for This Generation/Summary Chart of Unfulfilled Prophecy

15. A PROMISE TO REMEMBER   185

A Heartwarming Promise for Troubled Days/The New Promise Different From the Second Coming/The Rapture: An Important Teaching of the Apostles/The Rapture to Include Christians Who Have Died/Understanding the Order of Events/Christ's Authority to Order Resurrection/A Reunion in the Air/The Comfort of the Rapture/For the Living — Rapture and a New Body/The Rapture Must Occur Before Other End-Time Events/Living in Expectation

16. WHAT NEXT?   195

Watching for the Next Move/God's Perfect Timing/Why Has Christ Delayed His Coming?/The Clear Message of Hope/Are There Signs of the Lord's Return?/The Three Major Divisions of Prophecy/A Prophetic Checklist for the Church/A Prophetic Checklist for the Nations/A Prophetic Checklist for Israel/An Intricately Woven Pattern of Events/The Final Stage Is Set/How to Have a Happy and Meaningful Life

Preface

Each day's headlines raise new questions concerning what the future holds. The world has never before witnessed so many ominous developments and such rapid change. For the Christian, a natural response in these circumstances is to search the Bible to find what it has to say concerning the great events which lie ahead for the world.

The Bible contains the history of the world up to the present time, including its creation, the beginning of the human race, and God's dealings with Israel and the nations of old. The New Testament adds its important account of the birth, life, death, and resurrection of Jesus Christ in His first coming. The teachings of Christ and the apostles gave the early church a system of truth that not only explained the past but also presented a prophetic panorama of events yet to come.

Central in this scriptural revelation is a careful explanation of the second coming of Jesus Christ. Many predictions are made of the important events that will occur before Christ's second coming. When these events are placed in their proper order, the result is a prophetic calendar of what may soon overtake the world. An understanding of biblical prophecy has led many intelligent students of the Bible to believe that the world has already begun the countdown leading to

Armageddon, the greatest war of all history, and the end of civilization as we now know it.

Growing tensions in the Middle East and recent developments in the world have led many to ask whether we are moving rapidly to the end, to a gigantic, suicidal world war which many associate with the battle of Armageddon. Is the present world situation ripe for the fulfillment of the great prophecies leading to such a catastrophe? If so, what will happen next? What is the future of Israel? How will the energy crisis affect the future of the world? What are the events to watch in the countdown to Armageddon?

This book is an attempt to answer these important questions with special attention to the Middle East and the prophecies related to it. All agree that our world is entering a new series of one dangerous crisis after another. The days ahead promise rapid change and surprising developments. Only a word from God Himself can provide sure answers to important questions about the future.

The preparation of this book is the result of many years of specialized study in the field of biblical prophecy and has been preceded by a number of books related to this important area of biblical revelation. This volume is intended to answer normal questions about the future in plain language. While prophecy is an intricate subject, its main themes are understandable by anyone who is searching for answers.

Advanced students of the Bible, however, will find that this volume casts additional light upon important prophetic questions which face us in our modern world. In compiling this material, I have enjoyed the active editorial assistance of my son, Dr. John E. Walvoord, who has worked with me to produce a treatment of these great subjects in popular language. In quotations of the Bible, Old Testament passages are usually quoted from the New American Standard Bible (NASB) and New Testament passages from the New International

Version (NIV) and in some cases the Authorized Version (AV).

It is hoped that those who read this volume will be attracted to Jesus Christ as the only Savior and Lord and will have an intelligent understanding of how history is moving on to its climax. If the facts and conclusions of this volume are correct, the coming of the Lord for those who have trusted in Him may be expected momentarily.

JOHN F. WALVOORD

1

Questions About the Future

1

Questions About the Future

The Incredible Twentieth Century

By every standard of measurement, the twentieth century will go down in history as the incredible century. The century began with limited use of electricity, without radio, television, planes, missiles, electronic computers, modern weapons, and atomic bombs. Technology has moved man into a modern era which now witnesses more rapid change in the course of a year than formerly took place in a century.

Rapid Change

The phenomenal scientific developments of our age have exploded in parallel cultural changes. Gigantic manufacturing complexes have arisen; millions of people have moved from rural areas to the city; modern travel has shrunk the world so that everyone is everyone else's neighbor; luxuries, comforts, and pleasures, unknown in previous generations, have become commonplace. The electronic media has brought wars and disasters, world leaders and propagandists directly into our living rooms. Television transmission via satellite now makes it possible for all the world to experience swiftly changing events simultaneously. The shock of rapid change, in turn, has produced family and social crises, tensions between nations, fear of atomic war,

11

and struggle for survival in a world threatened by pollution, starvation, and moral disintegration.

The Impending Crisis

The rapidly increasing tempo of change in modern life has given the entire world a sense of impending crisis. Civilization is moving at a rapidly accelerating pace like a gigantic machine which will ultimately tear itself apart. How long can a world exist with atomic bombs unused, with increasing population and decreasing food supply, with growing moral degeneracy threatening to destroy our civilization? How long can world tensions be kept in check? How long can a world struggling for survival be kept from a bloodbath? The ingenuity of man has devised means of human destruction which would seem incredible to a previous generation. The world is moving toward a gigantic crisis. Can it be that the prophets of doom are right? Is the world racing toward Armageddon, a divine judgment on a wicked world, the end of human history?

The Voice of the Prophets

Prophecy has never been the sole domain of astrologers, mystics, and crystal-ball gazers. For centuries, people of all religions have considered the message of true prophets to be God's revelation of His plan for human history. The Old Testament books record the lives and works of many great prophets — Moses, Isaiah, Jeremiah, Ezekiel, Daniel, Hosea, Micah, and others. These men predicted future events in vivid detail, including the rise and fall of every major world empire which left its mark on the Middle East. Some of their predictions came true within their lifetimes, and many of their astounding predictions seem to be coming true today.

Jesus Himself claimed to be a prophet and quoted from Moses, Isaiah, Daniel, Jeremiah, and Micah — many times adding interpretive comments and detailed predictions of His own. His words have been confirmed by the test of time. Jesus' prediction of the fall of Jerusalem (Luke 21:20-24) was so vivid that the early church in Jerusalem was able to escape almost certain destruction by fleeing the city before Titus destroyed it in A.D. 70. As then, so now, crucial predictions about Jerusalem and the nation of Israel are, in fact, the key to understanding a carefully predicted chain of events which will mark the last days of our civilization.

The "Times" Are Running Out

Jesus predicted the persecution of the church, the fall of Jerusalem with the destruction of the temple, the scattering of the Jews into all nations, and the amazing survival and growth of the church. Along with the Old Testament prophets, He also saw a time when Israel would be reestablished as a nation. All this has been realized in history. But He also warned those who understood the Old Testament prophets to watch Jerusalem and the Middle East for signs of the approaching end of the world civilization, the end of the times of the Gentiles. The Jews, He said, ". . . will fall by the sword and will be taken as prisoners to all the nations. Jerusalem will be trampled on by the Gentiles until the times of the Gentiles are fulfilled" (Luke 21:24, NIV).

The Jewish people are now back in the land. Jerusalem, the city of dispute and negotiation, was won in 1967 only to become the object of a diplomatic tug-of-war. What is the future of the Holy City? If the times of the Gentiles are, in fact, nearing an end, what will happen next? Is there a sure word about tomorrow? Is it a word of hope or a word of doom?

The Search for Answers

Modern man is asking questions about the future as never before. They are solemn questions; they are searching questions. Only the Bible has clear answers. For this reason, biblical prophecy is being probed for clues to find where we are in God's program and what great events may be shortly coming to pass. Prophecies, formerly brushed aside as incredible, are now being studied again. Students of the Bible are becoming increasingly aware of a remarkable correspondence between the obvious trend of world events and what the Scriptures predicted centuries ago. People in all walks of life and of all religious faiths are asking the same question — What does the future hold?

2

Armageddon Calendar

The Valley of Megiddo, future site of the battle of Armageddon.

2

Armageddon Calendar

No Vague and General Predictions

Unlike the self-proclaimed prophets of today, the prophets of the Bible did not peddle vague and general predictions that could be adjusted to any situation. The prophecies recorded in the Bible are detailed and intricately interwoven. Although interpretation of minor points may vary, the overall picture is frighteningly clear. The Bible does not simply speak of a final destructive world war but of a whole series of carefully timed events on a doomsday calendar leading to Armageddon.

"Armageddon" has come to describe anyone's worst fear of the end of the world. The prophets have described it more specifically as the final suicidal battle of a desperate world struggle centered in the Middle East. This is the final act in a terrifying series of events which are very much related to today's headlines. This final, history-shattering battle will occur on schedule at a specific time and in its predicted location. The name "Armageddon" actually comes from a Hebrew word meaning "the Mount of Megiddo," a small mountain located in northern Palestine at the end of a broad valley. This valley has been the scene of many military conflicts in the past and will be the focal point of this great future conflict.

Seven Years Plus . . .

The final countdown involves years rather than days. Even before the countdown, several preliminary moves, which are predicted in the Bible, will shape the political, economic, and religious climate necessary for end-time events. These preliminary moves are now falling into place in rapid succession. As these moves are completed, a more specific timetable of events can begin. The final countdown involves a brief period of preparation plus seven more years — three-and-a-half years of comparative peace and three-and-a-half years of unparalleled disaster and war, climaxing at Armageddon.

Will Israel Survive?

A careful study of both history and the Bible is necessary to put the Armageddon calendar together. The reestablishment of the nation of Israel in the Middle East was the necessary start. Then came the series of Arab-Israeli wars, threatening to bring the United States and Russia into a direct clash over the Middle East. Prophecy seems to indicate that Israel will not be destroyed by war. Instead, Israel will eventually be betrayed and forced to accept an outside settlement at the peace table. The Arab oil blackmail and new economic and political alignments in the Middle East and Europe will eventually rob both the United States and Russia of a determining voice in the final settlement.

Oil and the New Power Grab

Dramatic realignment of political and economic power on the international scene is already in the making. For the immediate future the picture will look something like this: The power of Arab oil and European agriculture and industry may lead to a cartel that will eventually eclipse the power of both Russia and the United States in the Middle East. Stung by Vietnam and no longer needed by former European allies, the United States could easily sink into the mire of its

18

own economic problems and become isolated from world influence. Rejected by former Arab friends who absorbed billions in aid, Russia would be forced to wait on the sidelines. An emerging confederacy of Mediterranean nations will consolidate new economic and political power, regulating the rise and fall of national currencies, controlling world trade, and managing the bulk of the world's energy reserves.

The Modern Dark Ages

Belief in supernatural, mystical, and bizarre phenomena will grow from fad and superstition to a stampede of fear. Books, magazines, and movies about witchcraft, demons, and Satan-worship will soften the Western consciousness. Claims of demon possession and exorcisms will be increasingly in the news.

One factor which will have a traumatic effect on the world will be the fulfillment of what theologians call the rapture of the church, the sudden removal of every Christian from the world. This will fulfill the promise of Christ to His disciples when He said, "I will come back and take you to be with me that you also may be where I am" (John 14:3, NIV). At this time Christians who have died will be resurrected, and every true Christian living in the world will be suddenly removed to heaven without experiencing death. The disappearance of millions of Christians in a moment of time will deepen the religious confusion. The organized church with every true Christian removed will fall into the hands of political opportunists. Witchcraft, Satan-worship, and demonism will become increasingly prevalent throughout the world. World leaders will seek alliance with religious leaders to consolidate their powers in a modern culture as strange and mystical as the Dark Ages.

Armageddon Countdown

The first moves set the stage, but the precise countdown of seven years will begin with the signing of a final peace settlement in the Middle East. During the preliminary moves, the international balance of power will become more and more concentrated in a confederacy of Mediterranean and European nations. From the many negotiators and leaders involved in the Middle East one new international leader will emerge to superimpose a peace settlement on Israel and the more militant Arabs. This will bring an era of false peace, a move toward disarmament, and a major push for a new world economic system. These first three-and-a-half years will be the calm before the storm as the new leader of the Mediterranean consolidates his power.

The last three-and-a-half years will contain a series of almost inconceivable catastrophes. Just before this period begins, Russia will attempt a final bid for power for the Middle East, and her armies will be supernaturally destroyed. The balance of power will swing decisively to the new strong man controlling the Mediterranean Confederacy. As Satan's man of the hour, he will attempt to destroy Israel, now disarmed and at peace. In the fashion of the Babylonian and Roman emperors, he will deify himself and demand the worship of the world. Jews, professing Christians, and minorities of all kinds will be caught in a worldwide persecution that will exceed Hitler's methodical massacre of the Jews during World War II.

The world will begin to come apart at the seams — worse than any ecologist's nightmare. Acts of man and acts of God will combine to cause great disturbances in the world and in the solar system. Stars will fall and planets will run off course, causing chaotic changes in climate. Unnatural heat and cold, flooding, and other disasters will wipe out much of the food production of the world. Great famines will cause millions to perish. Strange new epidemics will sweep the world, killing millions in spite of all that modern medicine can do. As the period draws to a close, earthquakes will level the great cities of the world, and geographic upheavals will cause mountains and islands to disappear into the sea. Disaster after disaster will reduce world population in the course of a few years to a fraction of its present billions.

From World War III to Armageddon

Topping even these disasters will come a world war of unprecedented proportions. Hundreds of millions of men will be involved in a gigantic world power struggle centered in the Middle East. The area will become the scene of the greatest war of history. Great armies from the south, representing the millions of Africa, will pour into the battle arena. Other great armies from the north, representing Russia and Europe, will descend on Palestine. Climaxing the struggle will be millions of men from the Orient, led by Red China, who will cross the Euphrates River and join the fray. Locked in this deadly struggle, millions of men will perish in the greatest war of all history. This is what the Bible describes as Armageddon. Before the war is finally resolved and the victor determined, Jesus Christ will come back in power and glory from heaven. His coming, accompanied by millions of angels and saints, is described in graphic terms in Revelation 19. Coming as the King of kings and judge of the world, He will destroy the contending armies and bring in His own kingdom of peace and righteousness on earth.

From many indications, students of biblical prophecy have concluded that the world is well on the road to Armageddon. What are its signposts? How near is the world to the end? What is the hope of the individual in a time such as this? What can anyone in this generation do to survive? The answers to these pointed questions are found in the prophetic program of God dealing with the end of the age, as prophecies relating to Israel, the world, and the church are fulfilled. In the center of the stage is the little nation Israel, insignificant in number among the billions of the world's population and yet the fuse for the final world conflict which is ahead.

3

The Israeli-Arab Conflict

Israeli soldiers man positions in the Golan Heights.

3

The Israeli-Arab Conflict

The Slender Thread of Hope

Hundreds of years have passed since the destruction of Jerusalem and the slaughter of a million Jews by the Roman armies in A.D. 70. These centuries have been perilous times for all Jews. As Jesus and the prophets predicted, the Jews have been scattered throughout the nations of the world, victims of the sword of Rome, the inquisitions of a misguided church, and the gas chambers of Germany. But the prophets' words of doom were always mixed with the promise of a new day. From these promises came the slender thread of hope which seemed to hold the Jews together as they struggled to maintain their identity as a people.

Freedom — Land — Prosperity

Unless the prophets were idle dreamers, Israel's ultimate future was secure. According to the Scriptures the Jewish people would not only survive but also discover their freedom and possess their land as a new and prosperous nation in the Middle East. Critics of the Bible have laughed at these promises to Abraham and prophecies which run throughout the Pentateuch and the rest of the Old Testament.

The Jewish people from Abraham to the orthodox Jew of today have clung to these promises which have meant so much to Jews throughout the world. During the early war which threatened to exterminate Israel in 1956 and again in 1967, Radio Jerusalem broadcast the words of the Lord through the prophet Amos: "Also I will restore the captivity of My people Israel, and they will rebuild the ruined cities and live in them, they will also plant vineyards and drink their wine, and make gardens and eat their fruit. I will also plant them on their land, and they will not again be rooted out from their land which I have given them, says the LORD your God" (Amos 9:14, 15, NASB).

The prophecies to Israel were clear. A future time of rebuilding in the Promised Land was clearly predicted after which Israel would never again be uprooted from her land. Was the birth of the nation of Israel in 1948 the beginning of the fulfillment of that promise? If so, no matter what the odds, no nation or group of nations would be able to push Israel into the sea.

Arab Goal: Destroy Israel

From the standpoint of the Arab world, the presence of Israel in the Middle East is a festering sore that can only be remedied by radical surgery. Just as Israel has clung to its hope of a Promised Land, so the Arab world has clung to its hope of driving Israel into the Mediterranean and restoring the entire land, including Jerusalem, to Arab possession. To the Arab world, it seemed only a matter of time until its superior numbers and rapidly increasing wealth would make the survival of the nation of Israel impossible.

War of 1956: Arabs in Retreat

The history of the Israeli-Arab struggle from the beginning of the founding of the nation of Israel in 1948 has been one of increasing strength on both sides. In October 1956, the efficient armies of Israel easily overran the Gaza Strip. If Israel had not been stopped by action of the major powers, Egypt would soon have fallen to the advancing armies of Israel.

Russia Turns Against Israel

The years which followed were uneasy ones as Israel continued to have conflict with Egypt to the west, Syria to the north, and Jordan to the east. Under the leadership of Nasser, Egypt moved into closer alliance with the Soviet Union, which resulted in her receiving billions of dollars worth of military hardware. The Russian presence in the Middle East was painfully evident as thousands of Soviet military advisers poured into Egypt. Russian interference in the internal affairs of other Arab nations caused serious problems. Only strong United States and British intervention prevented direct Communist take-overs in the Middle East, as in July 1958, when United States Marines landed in Lebanon and British forces supported King Hussein in Jordan.

No Help From the United Nations

The Arab world, led by Nasser, continued to plot war against Israel. Nasser, under pressure from Syrian extremists, approved raids on Israel, even though it risked another war. Meanwhile, the United Nations obviously ignored the many border incidents designed to provoke Israel. When Secretary-General U Thant received a request from Israel to publish an official list of 100,000 border incidents since 1949, permission was refused because the list was too long to be practical. Israel was finally provoked to attack the little Jordanian town of Ses Samu and methodically destroyed it. While the Arab incidents did not provoke the United Nations to action, Israel's attack brought a formal United Nations condemnation. It became clear that Israel could expect no help from the U. N.

The Six Day War

Russia was doing all it could to capitalize on the situation. In 1967 she sent exaggerated reports to both Egypt and Syria about Israel's war preparations. Nasser believed the hour had come to destroy the nation of

Israel. The announced Arab goal was total war, total destruction, the death of every Israelite, and the reclamation of the land. One step led to another, with both sides receiving inflammatory information. Finally, the Egyptians blockaded the crucial Israel port of Eilat on the Gulf of Aqaba, making a confrontation inevitable.

Israel decided her only hope of survival was to attack first. The zero day was 5 June 1967. In the early dawn Israeli jets, coming in low from the north and northwest, destroyed the Egyptian air force. At the same time, Israeli torpedo boats and commandos demolished most of Egypt's naval power.

Israel's Miraculous Victory

The war which followed was devastating for the Arab world. Instead of exterminating Israel, Arab leaders saw their own armies and armaments ground to bits. Modern Israeli armor raced through the Sinai in a battle plan which drew heavily on strategies used in ancient battles in the same area, as recorded in the Bible and Jewish history. By the second day, it became evident to Russia that her plan had backfired. Three billion dollars worth of investment in military aid to the Arab countries was almost completely lost. Jordan and Egypt were already prostrate, and Syria was in retreat. It was only a matter of hours before Israeli troops could have begun assaults on Cairo, Amman, and Damascus.

Russia Attempts to Call a Cease-Fire

As in the war of 1956, as soon as Israel's victory became apparent, the cry went up for a cease-fire. At first, Russia opposed a cease-fire until it became evident it was the Arabs' only hope. The United Nations, which had done little to prevent the outbreak of the war, was now insisting Israel should stop. But Israel would stop only under certain conditions. Israel wanted recognition as a sovereign state and some permanent basis for a lasting peace.

The Arabs were still not willing to admit the facts, and the war continued, widening Israel's occupation of territory. With utmost care to prevent damage to religious shrines, the Israelis took Jerusalem. A high emotional pitch was reached in the war as the Jews for the first time took possession of the site of their ancient temple and the Wailing Wall. At the Wall hardened commandos broke down and cried for joy. General Dayan declared to the world press that the Jews had returned to their Holy of Holies, never to retreat again. The 1948 armistice had guaranteed their access to the Wailing Wall, but in the years that followed, the promise had not been honored.

Israel soon swept to the Jordan River, and the armies of Jordan were in retreat. Jordan's early defeat came partly because Egypt had not fulfilled its pledge of air support. Meanwhile, the Israeli army continued its advance against Syria and Egypt. In spite of Russian help in many ways, the battle was unequal.

Jordan ordered a cease-fire on Wednesday and was followed by Egypt on Thursday of that fateful week. Only Syria continued to struggle in spite of terrific odds. The Golan Heights, from which Syria had frequently bombarded Israel, were conquered by Israel at terrific cost. It was the beginning of the end for Syria, and in six short days the war was over.

Nasser's Myth of American and British Intervention

Nasser soon realized that his dream of conquering Israel had been shattered. To cover his devastating defeat, he accused Britain and America of joining in the struggle on Israel's side — a charge which was later proven to be totally false. On the concluding day of the war, Nasser, in a carefully rigged situation, announced his resignation. In spite of the terrible defeat of Egypt's army, the people clamored for Nasser's return. He continued to blame Egypt's defeat on British and American intervention — something which never took place.

Results of the War of 1967

As a result of the war, Israel increased her territory from eight thousand to thirty-four thousand square miles and doubled her population. Most important from the prophetic point of view, Jerusalem was back in the hands of Israel. The prospect of another war was averted for the time being. Israel had suffered less than a thousand battle fatalities in contrast to thirty-five thousand Arab dead. Israel had tremendously increased her stature as a nation among nations and had left the military might of her enemies in shambles. The miraculous war of 1967 was not to be the last, but the world had begun to notice the prophets' prediction that the Jews "will not again be rooted out from their land" (Amos 9:15, NASB).

Russia Restocks Egypt and Syria

In the years which followed, Russia again continued to arm Egypt, Syria, and other Arab countries. Considerable economic aid was given, and by 1972, twenty thousand Russian military personnel were in Egypt. Although Russia, at least outwardly, was attempting to hold down Egypt's warlike ambitions to conquer Israel, it seemed unmistakable that the long-range program involved just such a war. But Russia's growing military presence gave Egyptian leaders less and less control of their own military strategy and decisions.

Egypt Breaks With Russia

In July 1972, in a dramatic ninety-minute speech, Anwar Sadat, Egypt's president, announced that all Russian military advisers should leave the country and that Egypt would control all its own military bases and equipment on Egyptian soil. This setback to Soviet ambitions in the Middle East was only a taste of a new

Arab independence that would soon assert itself in the oil blackmail. The open disdain of the Russians toward the Egyptian Army and obvious manipulations for Russian control made Egyptians distrustful of Soviet ambitions. Egypt was convinced that Russia would never really equip Egyptians with sufficient first-class military weapons to win a decisive war with Israel. In this quarrel only part of the Russian soldiers left, and at least eleven thousand continued to man missile sites and defensive installations in Egypt. The open departure of thousands of Russians from Egypt was welcomed by the people who had little love for the Russians but Egypt was in a difficult situation. Without some new Arab strategy for continuing their objectives, they were forced to continue an uneasy dependence on the Russians.

Russia Repairs Break in Relations

On their part, the Russians had left quickly and had carefully pulled out all the sophisticated equipment not in permanent installations. Many of their long-term projects were left unfinished. The flow of spare parts suddenly slowed, and Egypt discovered her awesome dependence on Russia for spare parts, ammunition, technical advisers, and economic aid. The Arab countries had few friends to supply military equipment, and Egypt, at least, had no cash reserves to purchase such supplies. Meanwhile, Russia had begun to offer other Arab nations extensive supplies and economic aid. Superficially, at least, the damage in relations between Russia and Egypt was repaired. Military supplies once again flowed. Meanwhile, new alliances were being formed between Egypt and Saudi Arabia, and important discussions were being held to devise a new Arab strategy.

The Arab Oil Blackmail

Behind the new politics in the Middle East was a shift from reliance on Russia to an independent strategy in international diplomacy. The Arab nations began to threaten the industrialized nations of the world with a new weapon — soaring oil prices and an oil embargo against unfriendly nations. By September 1973 the pressure was already on the United States to withhold support from Israel as the price for a continued flow of oil. While this situation had been anticipated for years, it suddenly began to take a new and dangerous form. The need for oil was now complicated by the rapid increase in use of fuel throughout the world. In the United States, the shortage of oil and refining capacity was acute. Ecologists had succeeded in stopping progress of the Alaska pipeline, offshore exploration for oil, and the construction of nuclear power plants. Now lack of oil as a critical fuel was a worldwide threat. Libya, under its emotional leader Muammar Kaddafi, nationalized all its oil industry, giving itself control over exports of oil to the United States and Europe. The goal was to put pressure on the United States to withdraw its support of Israel.

War Preparation in 1973

Meanwhile, preludes to the conflict which erupted on Yom Kippur, the Jewish Day of Atonement, were already evident. Syrian MIG 21's challenged a flight of Israeli Mirages and Phantoms. The result was catastrophic for Syria, with thirteen of its planes lost and only one Israeli plane downed. This confrontation occurred as Arab leaders, including Syrian President Hafez Assad and Jordan's King Hussein, were gathered for a summit meeting with Egyptian President Anwar Sadat. Strange to say, the Arab Summit issued no ultimatums but was evidently busy repairing its internal alliances.

On 6 October 1973 the new war erupted as the streets of the cities of Israel were empty while people in the synagogues observed Yom Kippur. Suddenly, reserve units of Israel's army were called to active duty, and within hours air-raid sirens signalled that a war was underway.

Early Arab Victories

Egypt attacked the Sinai Peninsula by launching bridges across the Suez, with thousands of Egyptian soldiers and armor streaming across the bridges to occupy a large section of the Sinai Peninsula. At the same time Syrian armored forces attacked the Golan Heights, overrunning the few Israel defenders. The cease-fire which had closed the Middle East fighting in 1967 was in shambles. Another bloody war was underway. But this time it was different. The Arabs had attacked first, and Israel purposely had held back to make clear that they were the defenders, not the attackers. Militarily, this gave them considerable disadvantage and greatly increased the casualties.

World Reaction

The war took most of the world by surprise. It was assumed that Russia and the United States had agreed not to support a war in the Middle East, and even the Arabs had given indications of yielding to diplomacy rather than to military action. The objectives of the new war were not clear, as this time it seemed evident that the Arabs had no illusions of wiping Israel off the map. What they wanted was territory to be used as a basis for negotiation, a victory which would force Israel to make concessions. In this they almost succeeded. In one respect they were successful. The new unity Arab nations experienced during the war and the obvious economic clout of the oil blackmail began preparations for a new Arab plan. Their goal was to control the Middle East with a new diplomacy which would radically upset the balance of power in the world.

Mounting Casualties of the War

In the early days of the war Egypt was able to conquer a large portion of the Sinai Peninsula. Russia had furnished the Arabs with its latest sophisticated weapons. On the ground, Egyptian infantrymen using "Sagger" missiles were able to destroy Israeli tanks with one direct hit, and the weapon could be manned by a single soldier. The new Soviet SAM-6 missiles proved to have deadly accuracy in knocking Israel's planes out of the sky. Although some Egyptian planes were also hit in the confusion, the toll of Israeli men, tanks, and planes was tremendous. This was clearly a different war from 1967. The simultaneous fighting on two fronts against a heavily armored and fully equipped enemy taxed Israel's military capabilities. This time the Arabs did not flee at the first sign of conflict but in many cases proved themselves efficient soldiers.

The Tide Turns in Israel's Favor

As the war progressed, Israel's disciplined forces began to take their toll. Soon they had conquered a good portion of Syria and were in striking distance of Damascus. Exploiting a breakthrough at the Suez Canal, Israel began to sweep across the west shore of the Suez. Israel poured men and armor onto the western side of the Suez in an attempt to surround and isolate the Egyptian Third Army caught on the eastern side of the canal. Israel was soon in a position to cut off the Egyptian armies which had crossed the Suez into the Sinai Peninsula. Within days the missile sites on the western banks of the Suez could have been taken by the Israelis, which would have allowed the Israeli air force to pound the Third Army into submission.

Russia Presses for a Cease-Fire

Once again the Arab armies were on the defensive, and Russia again began to press for a cease-fire which would be to the advantage of the Arabs. When the war was finally stopped, twenty thousand Egyptian soldiers were pinned east of the Suez Canal with no route back to their homeland. Israeli military leaders were bitter about stopping the war only days before a clear victory on the Suez front.

Meanwhile, Russia and the United States both had been busy restocking their allies. Almost as soon as the war began, Russia started to restock Syria and Egypt. The United States countered with an equal supply of arms for Israel. By the end of the war, both sides had replaced all the equipment that had been lost.

New Realities in the Middle East

The war of 1973 made the world painfully aware that the entire planet was affected by the explosive Middle East situation. Any real peace would involve long and tedious negotiation with military incidents and the continued threat of renewed war. Radical Palestinian groups seemed determined to continue the violence and bloodshed until they were given a homeland in Palestine. Taken as a whole, however, the Middle East situation had moved into a serious phase. The United States was the sole support of Israel; Russia was committed to complete support of the Arab world; and the rest of the nations of the world were immobilized for fear of antagonizing the Arabs. This was just the beginning of new political and economic alignments which were much like those predicted by the prophets. The world was moving toward a dramatic realignment of nations similar to that predicted as leading to Armageddon.

The Miracle of Israel's Survival

In the center of the stage of the new Middle East situation is the continued presence and strength of the nation of Israel. The promises and prophecies about Israel to be fulfilled during her period of regathering and struggle to survive involve many details which are explained in chapter 5 of this book. One thing is clear — the last few decades of Israel's history have been miraculous, although deeply rooted in their ancient past.

The historical development of the Old Testament centers around Abraham and his descendants as a people chosen by God. They were heirs of the Promised

THE CHANGING MAP OF THE MIDDLE EAST

ISRAELI-OCCUPIED 1967 **1973**

Land and the prophesied source of the Messiah, the future King of kings and Savior of the people of God.

The Old Testament also contains sweeping prophecies concerning a period called "the times of the Gentiles." The New Testament confirms and elaborates these prophecies, explaining that the times of the Gentiles were to include not only the domination of history by non-Jewish nations but also the rise of the Christian church. During this period both Jew and non-Jew could participate equally in the spiritual blessings offered freely to all by the promised Messiah.

A Note of Warning

The emergence of Israel and the present struggle for international power in the Middle East may well be a note of warning that the end of the times of the Gentiles is only years away. The fact that Europe and Western nations have dominated history for the last thousand years has not set aside the promises to Israel and the promises of judgment and doom on all the nations who have afflicted her. The prophet Jeremiah gave this note of warning: "Therefore all who devour you shall be devoured; and all your adversaries, every one of them, shall go into captivity; and those who plunder you shall be for plunder, and all who prey upon you I will give for prey. For I will restore you to health and I will heal you of your wounds, declares the Lord, because they have called you an outcast, saying: It is Zion; no one cares for her" (Jer. 30:16, 17, NASB).

During a time of intense persecution of Jews and Christians, the Apostle Paul clearly indicated that the Old Testament promises for Israel were still to be fulfilled. Paul wrote, "I ask then, Did God reject his people? By no means!" (Rom. 11:1, NIV). The apostle then explained that the founding and success of the church during the times of the Gentiles was a mystery, not explained until New Testament times. But he warned his readers not to be ignorant, that this would not change the promises to Israel. What was happening to Israel would only continue "until the full number of the Gentiles has come in" (Rom. 11:25, NIV).

The Bible Proved True Again

Israel's future is carefully predicted by both the Old and New Testaments. The proof of the accuracy and truthfulness of the Bible is intertwined with the eventual fate of the nation of Israel. Against overwhelming odds, the Jewish people have survived. As Jeremiah prophesied, the enemies of the Jews have gone down one by one. The Roman empire was devoured and spoiled. The Russian czar who persecuted the Jews became a prey. The German Reich was destroyed. And now, miraculously, Israel is back in the land.

What will become of the Jews who have survived overwhelming odds, even extending their territory in spite of millions of Arabs determined to push them into the sea? What is the new and dangerous threat of the Arab oil blackmail? Will this new shake-up in international policy lead to the Mediterranean Confederacy predicted to dominate the world at the end time? Will tensions explode into the final desperate hours of the times of the Gentiles? Only biblical prophecies hold the answers to these important questions about the future.

4

The Arab Oil Blackmail

4

The Arab Oil Blackmail

New Arab Unity

In the aftermath of the war of 1967 in the Middle East, it soon became apparent that the relative military strength of Israel, Egypt, and Syria was only part of the total problem. Although the war did not end in victory for either Egypt or Syria, it indicated for the first time that with the help of adequate military support in personnel and equipment, the Arab countries would sooner or later have the military capacity to overwhelm Israel with her relatively small population. But would the rest of the world, especially the United States and Europe, allow Israel to be destroyed by war? Some new strategy was needed to make the Arab bloc independent of Russian support and free from possible intervention by a Russian-United States decision to preserve the status quo. If possible this strategy should also attempt to separate Israel from her allies, specifically the United States and Western Europe.

One of the important results of the war in 1973 was to bring the Arab nations together in a new unity which had never before been realized. Although the unity was manifested partially in the military conflict, it became increasingly evident that the Arab world's bid for power was going to be based on its control of the major oil resources of the world. In a new show of unity, the Arab world in November 1973 reduced its production of oil below the previous norm, and in the process attempted to embargo nations that favored Israel, principally the United States and the Netherlands. The grim prospect of having insufficient fuel for homes, industry, and military use sent shock waves around the world. A new war was in progress, an economic war of tremendous implications to the entire world. For the first time in centuries, the Middle East became a major component in every international consideration.

The World Energy Crisis

The rapid increase in consumption of energy throughout the world had been noted by experts for years, but warnings of an impending crisis fell largely on deaf ears. In industry, oil had proved to be cheaper and easier to use than coal. Much of the electrical generating capacity of the United States, especially in the East, depended upon energy derived from oil. In addition, ecologists had pointed out the air pollution caused by coal, particularly by the lower grades of coal which were economically feasible for industrial use.

The rapid increase of auto travel, especially in the United States, also presented a rising demand for oil for which there was no suitable substitute. Any reduction in oil would inevitably affect automobile travel and with it the total automobile industry which employed one out of every six people in the United States directly or indirectly. The prospect of limited supplies of gasoline threatened the life style as well as the economic prosperity of the United States. Europe and Japan were even more dependent on oil from outside sources, and the threat of limited oil produced panic unprecedented since the days of World War II. A whole new alignment of international power was underway, and no one could accurately predict the future.

66.4
Other

Middle East and
North Africa
390.9 Billion
Barrels

18.2
ABU DHABI

33.
IRAQ

62.2
IRAN

74.
KUWAIT

KNOWN WORLD
OIL RESERVES
WHEN THE
1973 WAR BEGAN*

137.1
SAUDI
ARABIA

Communist
Block
54.5

42.
RUSSIA

12.5 CHINA

North
America
47.2

8. CANADA

36.3
USA

2.9 OTHER

Other
65.1

10.3 EUROPE

15.3 CENTRAL AND
SOUTH AFRICA

14.
SOUTHEAST ASIA

25.5
SOUTH AMERICA

*PROVEN RESERVES AS OF 1973 EXCLUSIVE OF OIL SHALE

The Vital Role of Middle East Oil

On the basis of proved oil reserves, the Arab world was in a good position to control oil production and blackmail other nations. The Middle East and Northern Africa had approximately two-thirds of the known oil reserves of the world lying below their desert sands. Their known oil reserves stood at approximately 390 billion barrels as compared with all the other known oil reserves of the world at less than 170 billion barrels. Saudi Arabia alone had a potential of 137 billion barrels, almost as much as all the rest of the non-Arab world put together. Kuwait was next with 74 billion barrels, and Iran and Iraq were close behind with 62 billion barrels and 33 billion barrels. By comparison, the United States had 36 billion barrels and Russia, 42 billion barrels.

In addition these countries had sophisticated handling and processing facilities which had been developed by the major western oil companies. In rapid succession these facilities were nationalized. A few national leaders could now control not only a major portion of the world's oil reserves, but some of the most advanced production and processing facilities in the world. Without a continued supply of Middle East oil the industrial giants of the west simply did not have the energy needed for industrial production, transportation, electricity and heating.

Of all the nations of the world, the United States was the largest consumer, using one-third of the total oil production of the world. Still, the U.S. took no decisive action in the face of Arab threats. The U.S. managed to survive the 1973 embargo through a variety of indirect maneuvers including increased imports from Iran. But these tactics would provide no solution for threats of future embargoes or the continued high price of oil. As a non-Arab, but Moslem country, Iran soon strengthened ties with its Arab neighbors, especially Iraq and Saudi Arabia.

Oil at What Cost?

In the growing battle for world oil, the first step was to restrict oil output and delivery of Arab oil to the entire world. The cutback was linked to a demand for complete withdrawal of Israel from all lands which it had occupied by war, including the restoration of Jerusalem, or at least a portion of it, to the Arabs. Led by King Faisal of Saudi Arabia, the Arabs promised to increasingly restrict the shipment of oil to the entire world and embargo oil to selected nations like the United States. Along with the restrictions on the amount of oil delivered were shocking increases in the price of oil. Although the oil price had gradually risen from approximately $1 a barrel, suddenly oil prices began a spiraling rise, with some special sales going as high as $17 a barrel. This meant that instead of costing less than 3¢ a gallon at $1-a-barrel prices, crude oil would cost as much as 40¢ a gallon, making it almost prohibitive for wide use in supplying the energy needs of the world.

In short-term results, the Arab oil blackmail proved that Europe, Japan, and the smaller emerging industrial nations were more vulnerable to political manipulation than the United States and Russia. The impact in Europe was felt immediately as thermostats had to be turned down and automobile use was restricted. As the price of oil went up, so did almost everything else.

Japan was forced to pay out $18 billion for oil imports in 1974 — pushing the Japanese economy into decline and increasing inflation to a punishing 24%. Within a year inflation rates doubled in many countries — in Italy it increased to 25%, in Britain to 18%, in France and Belgium to 16%. Italy was forced to borrow more than $13 billion as the country tottered on the brink of bankruptcy. Great Britain braced itself for the worst economic days since the 1930's, trying to

45

muddle through until North Sea oil production could solve some of the country's economic woes.

But the price of Persian Gulf crude oil was not lowered after the crisis of 1973-1974. High oil prices and the continued pressure for price increases began to erode the economic foundation of the western world. The Arab oil blackmail had begun the fastest transfer of money in history. A few Middle Eastern countries were accumulating more wealth more quickly than even the Conquistadores when they seized the gold of the Incas. And the final price would not be exacted in gold alone, but in the political and economic reshaping of the world.

United States Oil Potential — Going It Alone

Immediately after the energy crisis in 1973 the entire world began with new determination to increase oil discovery and production. The rise in the price of oil suddenly made costly exploration and production methods worth the risk. But moves toward stimulating increased production or extensive exploration soon faltered in the U.S. as politicians debated and disagreed over energy policy decisions. Self-sufficiency seemed only a political slogan as the U.S. continued to play into the hands of the oil exporters.

In the United States the possibility of tapping large deposits of oil believed to lie off the coast of California, the Gulf of Mexico, Long Island, and Northern Florida was given reconsideration. Until the oil crisis, leasing in these areas had been strenuously opposed by residents who visualized their beautiful shores ruined by oil derricks with the possibility of oil spills and their devastating effect upon marine life and beauty. One of the immediate effects of the oil crisis was the approval of the Alaska Pipeline, which won early congressional approval. According to schedule oil would be flowing from Alaska at the rate of 600,000 barrels a day by 1978 increasing to about 2 million barrels a day in the early 1980's. But the Alaska Pipeline alone would not be of much help in releasing the U.S. from dependence on foreign oil.

Perhaps the largest known potential oil production in the U.S. has always been oil-bearing shale. Production of oil from shale deposits was not economically feasible or necessary until the Arabs increased the price of oil. A large formation of this shale, known as the Green River Formation, running through Colorado, Wyoming and Utah, had an estimated potential production of 600 billion barrels, almost twice the total Arab potential. More than 70 percent of these deposits were on government land, which could be leased to commercial oil producers. But two years after the Arab oil embargo no major pilot project of shale-oil production had been completed. The U.S. government had not leased enough of the land for a test program or provided private industry with price supports as an incentive for production.

In addition to developing oil-bearing shale the U.S. had hoped to develop vast amounts of synthetic oil and gas. A sobering study was released by the Federal Energy Research and Development Administration in 1975. The study warned that even under the best of circumstances it would take 25 years before the U.S. could expect much help in meeting its energy needs from oil-shale, synthetic oil and gas, or nuclear power plants.

Of all the major oil importers, the U.S. had seemed to be in the best position to develop self-sufficiency at the time of the Arab Oil Embargo. Yet, as time passed, the U.S. was unable to move forward with a decisive energy program. Oil imports began to increase in 1975. Late in 1975 the U.S. Geological Survey reported that recent estimates of U.S. oil reserves were considerably lower than anticipated. Future projections of oil needs estimated by private industry indicated that the United States would need increasing amounts of foreign oil until at least 1985. The U.S.

COST OF OIL IMPORTS

□ 1972 BEFORE THE OIL EMBARGO
■ 1974 AFTER THE OIL EMBARGO

BILLIONS OF DOLLARS

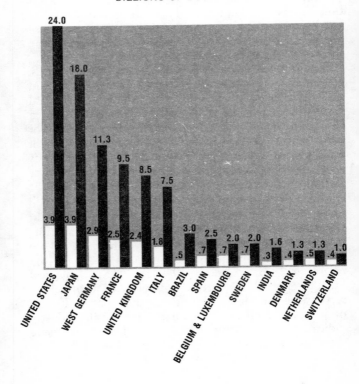

would need to import 10.6 million barrels of oil a day in 1985 — compared to the 6.3 million barrels of oil a day imported in 1975.

During the two years that followed the Arab Oil Embargo over 30 billion barrels of oil had been discovered and added to the proven reserves outside the control of the Organization of Petroleum Exporting Countries (OPEC). These major discoveries would begin to change the world picture in terms of oil production by 1985, but no sooner. Only England and Norway had any hope of real self-sufficiency in oil before 1985 — through the discovery and development of the North Sea Reserves. Enough time remained for a full-scale world disaster based on the demands of the Middle East oil producers.

U.S. Problems — Money, Time, and Ecology

The Western World had hoped that the U.S. would be able to solve its energy problems, even to help Israel and other nations that might be threatened by another oil embargo. But even after the Watergate Scandal, the new president and congress were deadlocked. Two years passed and precious time was lost. During that two years, government control of oil prices and other factors had caused U.S. oil production to decline.

In the total energy crisis, the U.S. was still in a favorable position with huge deposits of coal. Here, again, ecology became a major issue. Much of this coal could only be obtained by strip mining which destroys the beauty of the area being mined. A large percentage of the available coal was also of low grade, and its use would add greatly to air pollution. New methods would have to be required to avoid pollution if these low-grade coals were to be used for energy production. It was also possible to use coal to produce forms of oil but at a cost higher than previously considered feasible.

The U.S. was faced with a definite need to develop all of its energy resources in a crash program requiring the coordination of government and private industry. Valuable time had already been lost. The Federal Energy Research and Development Administration laid the plans for a program that would exceed the scope and expenditures of the Apollo program which spent $25 billion to send man to the moon. Even depending heavily on coal to get to 1985, the U.S. would still need vast amounts of oil — a good portion of which would have to be imported. Could the U.S. continue to pay for high priced imported oil and still spend billions on research? By the year 2000 solar energy and other sources could be developed. Would that be too late?

Faced with economic problems and years of inadequate planning for future energy needs, the U.S. must continue to depend on costly foreign oil. As long as this dependency exists the U.S. will not be able to play an important world role in determining what happens in the Middle East. Russia may soon be in the same position. Intelligence reports have indicated that even Russia, long an oil exporter, will be forced to import oil by the early 1980's. Although Russia has vast deposits of oil, production capacity is lagging behind demand. Soon Russia will also be shopping in the Middle East for oil. That leaves the most important future moves in the hands of a few Middle East rulers.

The Unhappy Jew

From the standpoint of the American Jew, the situation is decidedly an unhappy one. How long will America be willing to put up with sacrifices, fuel shortages, cold homes, limited automobile travel, and business disruption caused by fuel shortages just to help a small nation of a few million in the Middle East to survive? With the history of our withdrawal from Vietnam, where many more millions were involved and many American lives invested, is America in any

U.S. OIL NEEDS 1970-1990
IN MILLION BARRELS PER DAY PROJECTIONS

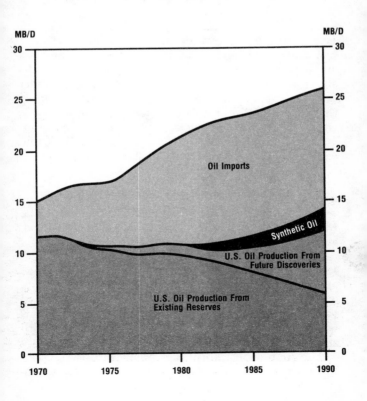

mood to sacrifice again for any country across the seas?

In the war of 1967 the Jew was a hero; but in the war of 1973 there were no victors. Even in Israel some believed that their own armies had not behaved well and that their own leaders had not been quick enough in their reply to aggression. Most uncomfortable, however, for Jews around the world was the danger of being blamed for the problems which beset the world as the Arab states reduced their supply of oil. Was there going to be another outbreak of anti-Semitism? Would the Jew once again be made the scapegoat as he was in Germany? Would America be willing to pay the price of being isolated, even from its European allies?

In this psychological and political struggle between Israel and the Arab world, Israel has found itself in a new series of dilemmas that have crossed political lines and have created new tensions and problems. The path ahead promises to be bitter, not only for Jews in Israel but for Jews around the world. For the leadership of the United States, a difficult choice is shaping up — either to force a destructive peace upon Israel, incurring the enmity of the Jewish population of the world, or to favor Israel and lose the friendship of industrialized nations around the world.

From Blackmail to Wealth and Power

The problem was one of colossal proportions. The Middle East, long the poorest area of the world, greatly under developed in industry and commerce, suddenly was in a position to dominate the world. Inevitably, major concessions would have to be made at the expense of the power of the U.S. and the security of Israel. New alliances between Middle Eastern and European leaders would eventually result in decisions which could drastically reshape the balance of wealth and power on this planet.

The Arab oil blackmail had begun with two powerful weapons — the oil embargo and staggering price increases in the cost of oil. The Arab oil embargo had attempted to control the flow of oil so that it would not reach countries friendly to Israel during the 1973 war. Since oil once sold was hard to trace, the Arabs discovered that it was impossible to boycott and isolate a specific country, as they attempted with the Netherlands. The total embargo of the western world was still useful as a threat, especially in the event of renewed fighting in the Middle East. But the strongest weapon in the Arabs' hands was control of the price of oil.

One measure of the success of the Arab oil blackmail could be calculated in dollars, petrodollars. The term petrodollars became a common way to describe the money paid by oil-consuming nations to the oil-exporting nations. Thirteen of the key exporting countries formed the Organization of Petroleum Exporting Countries (OPEC) and earned $100 billion in oil revenues in 1974. Saudi Arabia alone made almost $29 billion from oil sales in 1974. Iran made almost $21 billion, Libya almost $9 billion, Kuwait $8.5 billion, Iraq almost $8 billion, and the United Arab Emirates $6.5 billion. By the end of 1975 both the amount of oil produced and the price had gone up.

What was this to mean in terms of accumulated wealth? It meant that the nations who were forced to pay for oil at these higher prices would be stripped of their wealth, products and resources at the rate of over a hundred billion dollars a year. The surpluses that would be accumulated by the OPEC countries at this rate were estimated to be as high as $650 billion by 1978, but a more realistic figure would be $250 billion by 1978-1980. Even the lowest figure represents a massive amount of accumulated surplus wealth. In all its history the U.S. never accumulated more than $26 billion — the record in 1949. In 1975 the U.S. had only $16 billion in financial reserves and West Germany had $36 billion. By the end of 1975 Saudi Arabia had already surpassed both in accumulated surplus of financial reserves.

The Arab countries and Iran have achieved a new kind of power. They have oil, but they also have a powerful new relationship to western industrialized nations. They supply the oil, but they also supply the money needed for industrial investment and to finance government deficits. As purchasers of vast amounts of military hardware, machinery and technology they could choose to buy or not to buy from any particular country. In terms of money alone, the Arab countries have a vast amount of investable income that can be moved from bank to bank or country to country — which would play havoc with world currencies and world trade. The Arab countries could threaten to boycott not only oil but money. For political or economic reasons they could withhold money in the form of investment capital and the purchase of products and services.

In long-term results, the Arab countries and Iran have been able to consolidate both wealth and power. The cost to the world, and especially to Israel, would not just be calculated in petrodollars. These new realities were forcing a closer relationship between key European countries and some of the most powerful countries in the Middle East. Soon a group of no more than ten nations would be able to force a settlement in the Middle East and become the most dominant political force in the world.

End-Time Prophecy in the Making

The Bible predicted that a new alignment of nations, a Mediterranean Confederacy with great economic and political power, would emerge near the end of the times of the Gentiles. Until this time there has been neither the wealth nor the international need for such a shift in the balance of power. The high price of oil, which Europe's industrialized nations cannot continue to pay, has created the need for an economic cartel, a business arrangement in which the money spent on oil would be returned by the Arab bloc for investment, agri-

54

cultural goods and technology. The strength of such a cartel would quickly eclipse what either the United States or Russia would be able to accomplish during the next ten years. The political and economic implications to the world would be no less than those occurring when the Roman Empire combined many nations into one political and economic unity dominating the Mediterranean.

The most significant factor in the present Middle East situation is that the Middle East has risen to heights of world power such as have not been possible since the destruction of the Roman Empire. The prophetic significance of this rise in power in the Middle East is tremendous, and from a biblical standpoint it is the most dramatic evidence that the world scene is shaping up for Armageddon.

The United States and Russia in Decline

While the United States and Russia are the two greatest powers in the present world international scene, prophecy does not indicate that either will figure largely in end-time events. No specific prophecy whatever is found concerning the role of the United States, indicating that its contribution will be a secondary one as the world moves on to Armageddon. Russia, too, according to prophetic Scripture, is destined for declining power and influence. Instead, there will arise in the Middle East a new center of political and economic power, a new dramatic leader, and a new rapid sequence of events for which the present situation serves as a well-equipped stage.

The Prophetic Destiny of the Middle East

Civilization began in the Middle East with the creation of Adam and Eve. It was there that the great empires of the past rose and fell — Egypt, Assyria, Babylon, Medo-Persia, Greece, and Rome. It is in the Middle East that the future world government will have its center of political and economic power. The enigma of how the underdeveloped Middle East could ever become the center of world history again has suddenly been solved by the tremendous wealth and power latent in the oil reserves of the area. Already strategic as the hub of three continents, the Middle East is destined in the future to take a leading role in international and business affairs.

The Coming Crisis — How Far Away?

The coming crisis in the Middle East cannot be far away. The oil supply of the Middle East may run out in the next thirty to forty years. If the Middle East is to rise to power, now is its hour of opportunity. All signs point to an early confrontation as Palestinian Arabs continue to disrupt the Middle East and demand a state in Palestine. Whether there will be another war or not is uncertain, but Israel may be forced to launch a preemptive strike in a vain attempt to protect its interests. Arab oil producers can be expected to demand a settlement which would strip Israel of all the territory won since 1967 — including most or all of Jerusalem. The problem cannot continue unresolved for long. Arab strategists realize that the showdown must come soon or Arab unity will be lost. The new power of Arab oil and wealth can threaten restriction of oil production and economic reprisals to achieve its political goals. If the Arab world is to attain its goals, it must act quickly. Preliminary interim peace agreements such as Egypt and Israel signed in September 1975 do little to solve the basic problem.

5

Palestine:
The Land of Promise and Travail

5

Palestine:
The Land of Promise and Travail

The Horizontal View of Prophecy

The promises and prophecy pertaining to Israel are intricately interwoven throughout the entire Bible. Some involve general promises, but most involve specific and detailed glimpses into the future. When the prophets described future events, it was as if they were looking horizontally at distant mountain peaks. They described these peaks of history in vivid detail but often had little understanding of the vast valleys of time that separated the events they described.

Many of the prophecies about Israel are especially confusing unless they are carefully fitted together and compared to the history of the Jewish people and the Land of Promise. Three times the Jews have been scattered from the land — and three times they have returned. The prophecies of their scattering and return must be compared to the actual historical facts. Jerusalem has been destroyed many times. It is important to ask what prophecies have been fulfilled and which can be expected to be fulfilled in the future. Many of these ancient prophecies which have not yet taken place may well occur within our lifetime. An understanding of the prophecies in the Bible make it clear that Palestine's past, present, and future is a key to what will happen next in the world.

Prophetic Questions About the Promised Land

When the biblical promises and prophecies spoke of the Promised Land, did they describe an actual land with geographic boundaries? Was this literal or figurative language? If the promise of the land was an actual geographic location, have the Jews ever possessed it? Do they possess it or have a chance of possessing it today?

If the land was promised and prophesied as a reward for obedience, is it possible that disobedience or unbelief has caused the promise to be withdrawn? Is the promise restated in the New Testament? Are there conditions to that promise?

The answer to these important questions requires a further investigation into the specifics of the promises and prophecies about Israel. These specifics must then be compared to the actual history of the Jewish people and the Land of Promise.

Abram, the Man of Destiny

The twentieth century has witnessed one of the greatest miracles of world history — the restoration of an ancient people to the land of their fathers after centuries of being scattered among every nation of the world. The story of Israel's return, stirring as it is, has deep roots that reach back four thousand years to the city of Ur of the Chaldees, an ancient city in Mesopotamia located not far from Babylon. It was there that Abram, a man of destiny, lived in a city of advanced culture. To Abram came the command to leave the land of his fathers and go to a land God would show him. The story of Abram begins in Genesis 11:27 and ends with his death in Genesis 25:8. In a book which devotes only three chapters to describing the origin of the world, fourteen chapters are devoted to the life of Abram.

In partial obedience to God's command to go to a new land, Abram left his home and became a tent dweller, moving first a thousand miles to the northwest to Haran, where he lived until his father died. Then, another pilgrimage, this time another thousand miles to the southwest, brought him to the Land of Promise. It was there that Abram lived and died.

To Abram, man of destiny, God gave great promises. He promised to make him a great man (Gen. 12:2); He promised to make of his posterity a great nation (Gen. 12:2). Through Abram God promised that all families of the earth would be blessed (Gen. 12:3).

History has recorded the literal fulfillment of these promises. The name Abram, later changed to Abraham, meaning "father of nations," has been a great name, revered alike by Jew, Christian, and Muslim. Not only the nation of Israel, but the whole Arab world descended from Abraham. As God promised, his posterity indeed has become as the sands of the sea and the stars of the heaven — innumerable (Gen. 13:16; 15:5). Through Abraham came the prophets, the writers of the Scriptures of the Old and New Testaments, and Jesus Christ, the Son of Abraham, the Son of David. The nations that have blessed the descendants of Abraham have been blessed; the nations that have cursed the sons of Abraham have been cursed (Gen. 12:3). The judgments of God have fallen upon Israel's oppressors — Egypt, Assyria, Babylon, Rome, and, in more modern times, Spain, Germany, and Russia. The final judgments will come when the times of the Gentiles end.

The Promise of the Land

In the center of God's promises to Abraham are the promises to give his descendants a land. The whole drama of Abraham's life revolved around his willingness to leave Ur of the Chaldees to go to a new Land of Promise. The promise did not simply involve spiritual blessing; the lifelong pilgrimage of Abraham was a necessary part of the fulfillment of the promise for a land.

In Genesis 15 the confirmation of the promise is stated in no uncertain terms. Abraham's faith in leaving his homeland to find the Land of Promise was rewarded with a detailed explanation of the promise. The biblical narrative presents the description of the land in definite terms with clear geographic boundaries, and it even indicates the peoples who inhabited the land at that time: "On that day the LORD made a covenant with Abram, saying, To your descendants I have given this land, from the river of Egypt as far as the great river, the river Euphrates: the Kenite and the Kenizzite and the Kadmonite and the Hittite and the Perizzite and the Rephaim and the Amorite and the Canaanite and the Girgashite and the Jebusite" (Gen. 15:18-21, NASB).

This biblical account clearly referred to a specific geographical land, not some general blessing or a promise of heaven. The continued revelation of God to Abram elaborated the promise again as recorded in Genesis 17. Here Abram's name was changed to Abraham because of the promise that he would be the father of many nations.

The promise of the land is tied to the promise of a continuing heritage: "And I will establish My covenant between Me and you and your descendants after you throughout their generations for an everlasting covenant, to be God to you and to your descendants after you. And I will give to you, and to your descendants after you, the land of your sojournings, all the land of Canaan, for an everlasting possession; and I will be their God" (Gen. 17:7, 8, NASB).

The meaning to Abraham and the early readers of Scripture is clear. The land promised was Palestine, stretching from the Sinai Desert north and east to the Euphrates River; this would include all the holdings of present-day Israel, Lebanon, and Jordan, plus substantial portions of Syria, Iraq, and Saudi Arabia. What is even more interesting is that by the time of Genesis 17 the ultimate outcome of the covenant was clearly not conditioned on later obedience. The covenant was described as an everlasting covenant, and the land of Canaan promised as an everlasting possession.

Delayed Possession of the Land

But Abraham never possessed the land. He was one hundred years old when God miraculously enabled him to have a son, Isaac, to fulfill His promises. But even after the birth of Isaac, Abraham did not possess the land, except for the small portion which he bought as a burial plot for Sarah. Instead, God plainly told him that his posterity for hundreds of years would not possess the land; they would, in fact, leave the Land of Promise and go to a land that was not theirs (Gen. 15:13), and only later would they return to possess the land.

The First Departure From the Land

History has recorded how the children of Israel went down to the land of Egypt in the time of Joseph. Sold by his brothers into slavery, Joseph was eventually exalted to a high position of authority because of his interpretation of a prophetic dream of Pharaoh. Because of his position, Joseph was able to provide for his relatives and they were welcome in Egypt. In due time his father, Jacob, the grandson of Abraham, and his brothers and their families came to Egypt to survive the years of famine. There Israel grew from a large family of seventy to a nation of two or three million.

Israel in Egypt

Meanwhile, there had been a change of rule in Egypt, and Israel as an expanding alien population was no longer welcome. They found themselves in affliction as slaves of a cruel Pharaoh who sought their extermination and began a plan of killing every male child shortly after birth. In that hour of Israel's crisis, Moses was born, adopted by the daughter of Pharaoh, educated in the palaces of Egypt, and prepared to be Israel's deliverer.

The First Return

The book of Exodus records the first return to the land. This return was marred by forty years of wandering in the wilderness because of Israel's unbelief. Finally successful, however, Israel first conquered the land to the east of the Jordan. During this first return, the explicit promise of a land for the Jews was repeated to Joshua after the death of Moses. The command from God was to go over the Jordan "to the land which I am giving to them, to the sons of Israel" (Josh. 1:2, NASB). The geographic description of the land was as specific as the references in Genesis: "From the wilderness and this Lebanon even as far as the great river, the river Euphrates, all the land of the Hittites, and as far as the Great Sea toward the setting of the sun, will be your territory" (Josh. 1:4, NASB).

Finally, after years of wandering, the Jews crossed the Jordan, which was miraculously dried up for their passage. Then they conquered Jericho, Ai, and a large portion of the Promised Land. The hope of Israel had been partially fulfilled, and the promise had been clearly restated. Yet the early conquest never reached into all the land that had been promised.

Immediate Occupation Conditioned on Obedience

The children of Israel were warned that unless they kept the law which had been delivered to them in the wilderness, they would be driven out of the land again. Israel survived the moral degeneration during the time of the judges, rising to the glory of the kingdoms under Saul, David, and Solomon. The kingdom of Israel under Solomon controlled most of the land God had promised Abraham, stretching from the river of Egypt, the boundary of Egypt, to the river Euphrates far to the east. But the land was never fully possessed by the Jewish people, and this was clearly not the time the land was to be claimed and kept as an "everlasting possession." Israel's momentary victory was soon overshadowed by moral and political disaster.

The decline of the kingdom was marked by the popularity of pagan religions and moral degeneration. King Solomon had married women from surrounding pagan cultures, who raised his children to be idol worshipers instead of worshipers of the true God of Israel. The kingdom began to disintegrate from within. After Solomon's death it was divided into the kingdom of Israel in the north, composed of ten tribes, and the kingdom of Judah in the south, with the two tribes of Judah and Benjamin.

The Second Departure: The Captivities

The prophets had warned of impending judgment, and in due time it came. The Assyrians swept the ten tribes into captivity in 721 B.C. A century later, Nebuchadnezzar, the general of Babylon's army and soon to become king of Babylon through the death of his father, conquered Jerusalem in the late summer of 605 B.C. This invasion eventually resulted in the capture and transport of many of the children of Israel to Babylon. Jerusalem was finally devastated in 586 B.C. and left in shambles. Once again the people of Israel were out of the land, and Jerusalem was desolate.

The Promise of the Second Return

But the prophet Jeremiah, living in the times of Israel's downfall, predicted that Israel would return. Although the prophet had predicted the Babylonian captivity, he also predicted, "For thus says the LORD, When seventy years have been completed for Babylon, I will visit you and fulfill My good word to you, to bring you back to this place" (Jer. 29:10, NASB).

The Prayer of Daniel: The Second Return

Almost seventy years later the prophet Daniel, then an old man of eighty years of age, read the prophecy of Jeremiah (see Daniel 9). Daniel immediately prepared his heart for an extended time of prayer and intercession on behalf of his people. He claimed the promise of Jeremiah and asked God to restore His ancient city, left in ruins by the Babylonians, and once again establish it as the city of God and the capital of Israel. Daniel was an unusual man of prayer who was accustomed to praying three times a day with his windows open to Jerusalem, pleading with God to restore His ancient people and His ancient city (Dan. 6:10).

The Temple Rebuilt

Daniel's prayers were answered, and Jeremiah's prophecy was literally fulfilled. The book of Ezra records the story of fifty thousand people returning to the Promised Land. After considerable struggle, they finally rebuilt the temple in 516 B.C. The city itself was not rebuilt until after 445 B.C. Nehemiah came to survey the ruins of Jerusalem, with authority from the king to rebuild the city. At that time the walls of Jerusalem were once again constructed. In the fifty years that followed, Jerusalem was reestablished as the center of Israel's political, religious, and national life.

The Temple in the First Century A.D.

The years passed. In 20 B.C., the Roman government, on behalf of Israel, began a reconstruction of the temple which had been built in 516 B.C. The temple was almost five hundred years old and in a sad state of disrepair. This reconstruction project was in progress when Christ was born and was still unfinished when Christ died on the cross. It was not until A.D. 64 that the magnificent temple, constructed of stone from the quarries underneath the city of Jerusalem, was finally completed.

Destruction of the Temple

The glory of the temple, however, was short-lived. Six years later, in A.D. 70, Jerusalem was once again caught in the throes of war, surrounded by Roman soldiers who slaughtered hundreds of thousands of Jewish pilgrims who had returned for the feast. Ultimately, the city was destroyed and the stones of the temple were pried apart one by one and cast in the valley at the southeast of Jerusalem. Today, some of these same stones have been discovered by archeologists.

The Third Departure

With the slaughter in Jerusalem and the destruction of the temple, Israel's control of the Promised Land was once again shattered. In the years which followed, more cities were destroyed and Israelites were physically driven from the land and scattered all over the world. All this had been anticipated by Moses who had warned Israel in stern words about the consequences of disobedience to their law. "And it shall come about that as the LORD delighted over you to prosper you, and multiply you, so the LORD will delight over you to make you perish and destroy you; and you shall be torn from the land where you are entering to possess it. Moreover, the LORD will scatter you among all peoples, from one end of the earth to the other end of the earth, and there you shall serve other gods, wood and stone, which you or your fathers have not known" (Deut. 28:63, 64, NASB). This ancient prophecy continues to describe Israel's persecution, uncertainty of life, fear, and insecurity among the nations of the world.

Prophetic Hope for a Last Return

Jesus had clearly predicted that the magnificent temple of His lifetime would be destroyed stone by stone (Matt. 24:2). The fate of Jerusalem and the Jewish people was understood to take a drastic turn for the worse within the lifetime of the generation which heard His message. Was Israel finished? Were the promises to be set aside?

The Apostle Paul, himself an Israelite, struggled with these very questions. His conclusion was simply an emphatic no! "Did God reject his people? By no means!" (Rom. 11:1, NIV).

Would there be a fulfillment of the everlasting covenant to possess the land, a land with clear geographic boundaries promised as an everlasting possession? The writer of the book of Hebrews reviewed the promises to Abraham. Were they considered to be conditioned on Israel's obedience, now to be set aside because of Israel's disbelief? No, they were not. The final outcome was immutable, unchangeable. "When God made his promise to Abraham, since there was no one greater for him to swear by, he swore by himself , . . . Because God wanted to make the unchanging nature of his purpose very clear to the heirs of what was promised, he confirmed it with an oath. God did this so that, by two unchangeable things in which it is impossible for God to lie, we who have fled to take hold of the hope offered to us may be greatly encouraged. We have this hope as an anchor for the soul, firm and secure" (Heb. 6:13, 17, 18, 19a, NIV).

The clear and straightforward testimony of both the Old and New Testaments points to a final return to the Land of Promise. This hope of restoration has sustained the Jews through nineteen hundred years of struggle. Christians who have understood the clear prophetic witness of the Bible have expected it to happen for hundreds of years, even though the odds were overwhelmingly against it.

Scattered Israel's Persecutions

The centuries that followed were tragic for Jews. Afflicted by persecutions of incredible severity, their total number at one time shrank to approximately one million. Hated by professing Christians and non-Christians alike, they were driven from land to land, never finding rest or prosperity except for brief periods of time.

From all outward appearances, Israel's future was hopeless. In the eighth century the Arabs took possession of Israel's ancient land. In the twelfth century for a brief time the Christian Crusaders were established in Palestine, but were finally defeated by Saladin in 1187. The Ottoman Turks took over in 1517 and continued their control over the Promised Land until Turkey was defeated in World War I. A dramatic turn of events began in 1917 when General Allenby conquered Jerusalem, and the British occupation of Palestine began.

Beginnings of the Third Return

The return of Israel to the land was beset by many problems, although as early as 1871 a few Jews had managed to return to the land, and about twenty-five thousand Jews had settled there by 1881. The idea of the Jews repossessing their ancient land was not stated in any formal way until it was adopted in the first Zionist congress, called in 1897 by Theodore Herzl. The goal of reclaiming the land of Palestine as a home for the wandering Jews, although seemingly an idealistic dream, brought the light of hope to Jewish eyes around the world.

The Balfour Declaration, 1917

The progress was painfully slow, however. By 1914 the number of Jews in the land had reached only eighty thousand. During World War I, to gain support of Jews for the war effort, the British Foreign Secretary, Arthur J. Balfour, issued the Balfour Declaration on November 2, 1917. This declaration indicated approval of Israel's goal in the words, "His Majesty's Government views with favor the establishment in Palestine of a national home for the Jewish people . . . ." Pressure from the Arab world, which would have no part in establishing a home for Israel, and the desire of Britain to maintain friendship with the Arab nations prevented any fulfillment of this promise. Little progress had been made when World War II broke out in 1939. By that time, however, 400,000 Jews had managed to find their way into the country in spite of severe restrictions on Jewish immigration and laws that did not allow Jews to possess real estate.

The State of Israel Established in 1948

World War II, which recorded the terrible destruction of millions of Jews under Nazi tyranny, created a favorable attitude and worldwide sympathy for the Jewish people. Certainly there should be some place where the wandering Jew could have his own home. Although an Arab League was formed in 1945 to oppose Jewish expansion, the United Nations created a home for Jews in Palestine after World War II with the approval of the United States and Russia. The directive stated that Palestine should be divided into a Jewish state and an Arab state, and five thousand square miles were assigned to the infant state of Israel.

Israeli-Arab War, 1948-1949

On May 14, 1948, when British control ended, Israel for the first time in hundreds of years became an independent state. The population of the new nation included only 650,000 Jews and many hundreds of thousands of Arabs. The settlement, however, was quite unacceptable to the Arab world. Israel was attacked on all sides by Egypt, Jordan, Iraq, Syria, Lebanon, and Saudi Arabia. Israel's defense was heroic, and the armies of its opponents were disorganized. Although thousands of Israelites fell in battle, by the time a truce was arranged on 7 January 1949, Israel had extended its area of possession from five thousand square miles to eight thousand square miles, including much of the Negev, the great desert to the south. The history of the nation from that date to this has been one of unending trouble, recurring warfare, but gradual strengthening of the nation.

Have the Prophecies Been Fulfilled?

From the time of the promise to Abraham to the present hour, the prophecies about Israel's total possession and blessing in the land have not yet been fulfilled. The pulse of history, with three successive departures and three returns, has been a dramatic one. The most dramatic events, however, are still ahead.

Is the present third return of Israel the beginning of the last and ultimate regathering of Israel? Is this one more in a series of current events preparing the way for the end of the times of the Gentiles? As non-Jewish nations jockey into position for their last moments of glory and judgment, it is important to realize that, after the times of the Gentiles end, Israel will finally experience all that has been promised and hoped for since the time of Abraham.

No Easy Road to Glory

The final return of the Jews to Palestine will be no easy road to glory. Those who believe present negotiations in the Middle East will lead to a lasting peace will be sadly mistaken. The doves and peacemakers in Israel will eventually lead their country into a peace which will compromise their security in the hope of economic prosperity. It is sad to forecast that a nation that defended itself so bravely for almost thirty years will be lulled into a false peace with international guarantees, which will be of no help in the final hour the Bible describes as the "time of Jacob's distress" (Jer. 30:7, NASB).

71

After their initial return, the Jews will go through a period of false peace, an international betrayal, and a bloodbath of astounding proportions. Without the direct intervention of God, both Palestine and the Jewish people would be completely destroyed. The detailed prophecies revealing the fate of Jerusalem and the land are the subject of the next two chapters.

Jeremiah's prophecy describes the time of Jacob's distress in vivid terms. This is the last three-and-a-half years leading to the final battle of Armageddon. It will begin when the false peace with Israel is broken by the new leader of the Mediterranean Confederacy. This period of history is often called "the great tribulation," and is detailed in the book of Revelation as a specific period ending the times of the Gentiles. Palestine, Jerusalem, and the temple rebuilt during the peace will be ". . . given to the Gentiles. They will trample on the holy city for 42 months" (Rev. 11:2, NIV).

After the promise of the return, Jeremiah warned, "For thus says the LORD, I have heard a sound of terror, of dread, and there is no peace. Ask now, and see, if a male can give birth. Why do I see every man with his hands on his loins, as a woman in childbirth? And why have all faces turned pale? Alas! for that day is great, there is none like it; and it is the time of Jacob's distress, but he will be saved from it. And it shall come on that day, declares the LORD of hosts, that I will break his yoke from off their neck, and will tear off their bonds; and strangers shall no longer make them their slaves. But they shall serve the LORD their God, and David their king, whom I will raise up for them" (Jer. 30:5-9, NASB).

A New Kingdom Completes the Return

The final return will not be completed until Christ returns to intervene as Judge of the nations and the promised King to restore David's throne. The prophecies concerning the return, so dramatically begun in our history, will not be completed until all living Jews (drastically reduced in number during the great tribulation) are finally established in Palestine for a millennium of peace and prosperity. Many of the Old Testament prophets echoed similar visions of this period.

Jeremiah prophesied the judgment that would fall on the non-Jewish nations and then reported the message from the God of Israel in these words: "Then I myself shall gather the remnant of My flock out of all the countries where I have driven them and shall bring them back to their pasture; and they will be fruitful and multiply. I shall also raise up shepherds over them and they will not be afraid any longer, nor be terrified, nor will any be missing, declares the LORD. Behold, the days are coming, declares the LORD, when I shall raise up for David a righteous Branch; and He will reign as king and act wisely and do justice and righteousness in the land. In His days Judah will be saved, and Israel will dwell securely; and this is His name by which He will be called, the LORD our righteousness" (Jer. 23:3-6, NASB).

The same theme of a final return to be completed only after a period of judgment is recorded by the prophet Ezekiel. This prophecy makes it clear that this is finally accomplished directly after a period of judgment and that the completion of the return will bring every living Jew into the land. ". . . I made them go into exile among the nations, and then gathered them again in their own land; and I will leave none of them there any longer" (Ezek. 39:28, NASB).

Setting the Stage

The present return of Israel to the land sets the stage for an important series of end-time events. Prophecies have clearly predicted that Israel would be reestablished as a nation before the end of the times of the Gentiles. But non-Jewish nations were described as still in control of the destiny of Israel during a final hour of judgment and suffering. Jesus predicted that "Jerusalem will be trampled on by the Gentiles until the times of the Gentiles are fulfilled" (Luke 21:24, NIV).

The next dramatic moves will involve the struggle for control of Jerusalem and the return of a portion of the land to the Palestinian Arabs. This struggle will become increasingly explosive because of the three million dispossessed Palestinians. Many have been forced to live as refugees and exiles in surrounding Arab states. Palestinian guerrilla groups spawned from refugee camps have pursued a strategy of terror and have made it difficult for moderate Arabs to negotiate compromise solutions.

Arab leaders have managed an uneasy alliance with the Palestinians that has often exploded in violence. In October 1974 the Arab summit conference at Rabat offered a solution for the sake of unity. The occupied West Bank of the Jordan was to become an independent Palestinian state run by the Palestine Liberation Organization (P.L.O.). A government in exile was to be formed by early 1975 to prepare for negotiation with Israel.

Yasir Arafat, guerilla chieftain and leader of the P.L.O., was welcomed to address the United Nations and treated as a head of state. But Palestinians soon discovered they were better at terror tactics than developing a unified government in exile. Extremists continued to reject any coexistence with Israel and opposed any permanent settlement. No unified government in exile seemed possible and guerrilla raids were soon resumed against Israel.

The stage is set for an open struggle over Palestine and the city of Jerusalem. Guerilla raids and reprisals could easily push the Middle East into a new war. As moderate Arabs and European allies consolidate wealth and power a forced settlement will seem the only solution. New leaders will emerge to promise just such a peace. This hoax of peace and prosperity may lead to the formation of the most vicious world empire of all time.

74

6

The City of the Prophets

View of the temple area in Jerusalem, site holy to two re-
gions. The Wailing Wall is in the foreground and the Do
of the Rock is in the background.

6

The City of the Prophets

Hallowed Memories

As the world races toward its final hour of struggle, the important city to watch is not New York, Moscow, Paris, Peking, or Cairo. The city to watch is Jerusalem! Hallowed memories haunt its ancient streets and holy places. The stones and debris from the invasions of the past, upon which the modern city of Jerusalem is built, are fearful reminders of the prospect of future wars. But there is hope — hope of a glorious future for Israel. Etched in Jewish memories of the golden days of David and Solomon, the message of the prophets still echoes in the holy city. Their message of coming sorrow and worldwide dispersal of the Jews was tragically fulfilled in history and is reenacted in the tears still being shed at the Wailing Wall. Jesus also added His careful and detailed predictions about the future of the city and wept with compassion over what He saw was yet to come. These prophecies of the Old and New Testaments provide the final key to what will soon occur in the future dramatic unfolding of world history.

The City of Destiny

Revered by Christians, Jews, and Arabs alike, Jerusalem is already the center of religious interest for much of the world. Jesus and the prophets predicted this would be the case at the end time. In the light of these prophecies, it is understandable that the attempts for a permanent peace in the Middle East will involve claims and counterclaims regarding Jerusalem. Political, economic, racial, and religious issues are interwoven to make the control of Jerusalem the most explosive issue in seeking a peace in the Middle East. The real issue is neither the Sinai Peninsula nor the Golan Heights, though they frequently have been in the spotlight.

The conflict over Jerusalem will continue to be a potential powder keg, threatening war which could spread like wildfire from the Middle East to the entire world. Only an international peace settlement, perhaps forced upon the contending parties, can subdue the fires of conflicting opinion. Increasingly there will be a religious element in the controversy, as the religions of the Middle East struggle for a secure holy place in Jerusalem. Out of this will come Arab demands for at least a share in the holy ground where once the temple stood. Prophecy indicates a future attempt to rebuild Israel's ancient temple. In attempting to understand the Middle East situation, both history and prophecy must be taken into consideration in any interpretation of the direction of current events. Jerusalem, the city of the prophets, is the key to understanding the great predictions of the prophet Daniel in the Old Testament and of Jesus in the New Testament. Today Jerusalem is already moving toward its prophetic destiny as the center of dramatic end-time events.

The Ancient History of Jerusalem

The founding of Jerusalem is shrouded in the mysteries of the past. First mentioned in Genesis 14:18 as the city of Salem, over which Melchizedek was king, its early existence is attested by the Tell el Amarna Tablets discovered in Egypt. The city named Salem, later combined with the name "Jeru," has had a history of almost four thousand years. Jerusalem through the centuries has been better known by more people in more generations than the name of any other city in the world.

The Center of Israel's Life and Hope

Situated in the Judean hills, about halfway between the Mediterranean coast and the Jordan River, Jerusalem has been the scene of countless stirring events of the past. The city's rise to prominence began in the tenth century when David, the king of Israel, made Jerusalem his capital. The glorious reigns of David and Solomon brought splendor and wealth to the city. The crowning achievement of that great era was the construction of Solomon's magnificent temple in Jerusalem. It was to this holy city, in the ebb and flow of religious revival and decline, that the prophets of God came to deliver their messages of thundering denunciation coupled with glowing predictions of a glorious future.

Again and again the tides of war swept over this ancient city. Many times it was completely destroyed only to rise from its ashes to be born anew. Its conquerors have included armies from all the great nations of the past. The streets of Jerusalem have been trampled under the feet of Egyptians, Assyrians, Babylonians, Persians, Seleucids, Romans, Christian Crusaders, Saracens, Mamelukes, and Ottomans. Few cities of the world have had more dramatic and poignant memories.

The Messiah to Come to Jerusalem

For centuries in Israel's history, the Jews had awaited the coming of their Messiah. Their forefather, Abraham, from whom had come both Israel and the Arabs, had been given great promises. The land would belong to the sons of Jacob, the grandson of Abraham. But their possession of the land would not be immediate. They would go off to a distant land for hundreds of years and then return. This was fulfilled when Jacob and his family moved to Egypt and multiplied until they became a great nation of several million in number.

The saga of their exodus from Egypt and their possession of a portion of the land is contained in the early books of the Old Testament. But the prophets also predicted they would be driven out of the land again because of their failure to keep the law which God had given them. In due time the Assyrians carried off ten of the twelve tribes, and 125 years later the Babylonians carried off the two remaining tribes, leaving Jerusalem in ruins. But the prophets predicted their return, and back to Jerusalem they came in the sixth century B.C. Once again the temple and the city were built, and Jerusalem again became the center of Israel's religious and political life. Devout Jews understood from the Old Testament prophets that their Messiah, the one chosen of God to be their Savior and leader, would come to Jerusalem.

When the Holy Land fell to the disciplined Roman armies in 63 B.C., it began centuries of oppression and persecution for the people of Israel. The yoke of the Roman occupation was a heavy burden. The religious life of the people became obscured. Their religious leaders were divided, but they still longed for the coming of their Messiah and King who, according to the prophets, would rule on David's throne and deliver His people from oppression.

As recorded in the New Testament gospels, Jesus was born in the little town of Bethlehem, south of Jerusalem, just as the prophet Micah had said (Mic. 5:2). Mary, the mother of Jesus, and her husband, Joseph, brought the infant to the temple in Jerusalem according to the Jewish custom and presented the sacrifices required by the law of Moses. Spending His boyhood in Nazareth, Jesus returned to the temple in Jerusalem with His family for the Passover at the age of twelve.

There He astounded the scholars and religious leaders of His day with His comprehension of theology. After this brief episode, however, He returned to Nazareth and lived in obscurity until approximately twenty years later when He began His public prophetic ministry.

The Unwelcome King

Introduced by the prophet John the Baptist as the one of whom the prophets had spoken, Jesus immediately attracted the attention of the multitudes. Especially in Jerusalem, the center of Israel's religious life, the high priests and leaders were skeptical of Jesus' claims. But the common people followed Him by the thousands, impressed by His teaching and astounded by His remarkable miracles. A confrontation between Jesus and the leaders of Israel was inevitable. After three years of itinerant preaching by Jesus, the conflict between Jesus and the religious leaders of Israel broke into open and violent confrontation.

In keeping with Old Testament prophecy, it was at this time that Jesus directed His disciples to obtain a colt upon which He could ride for His entry into Jerusalem. Centuries earlier Zechariah had prophesied, "Rejoice greatly, O daughter of Zion! Shout in triumph, O daughter of Jerusalem! Behold, your king is coming to you; He is just and endowed with salvation, humble and mounted on a donkey, even on a colt, the foal of a donkey" (Zech. 9:9, NASB).

Just as Zechariah had predicted, Jesus entered triumphantly into Jerusalem. The people who had followed Him from outside Jerusalem and others from the city hailed Him as the predicted Messiah and Son of David who would rule over Israel. The gospel of Matthew describes the event in dramatic terms: "The crowds that went ahead of him and those that followed shouted, Hosanna to the Son of David! Blessed is he who comes in the name of the Lord! Hosanna in the highest! When Jesus entered Jerusalem, the whole

81

city was stirred and asked, Who is this? The crowds answered, This is Jesus, the prophet from Nazareth in Galilee" (Matt. 21:9-11, NIV).

The enthusiastic reception of Jesus as the Prophet and Messiah of Israel shocked the religious leaders. They demanded that Jesus silence His disciples and those who proclaimed Him as king. In this situation Jesus' answer was one of rebuke to these leaders. He said, "If they keep quiet, the stones will cry out" (Luke 19:40, NIV).

Jesus Denounces the Religious Leaders of His People

The hostility of the religious leaders toward Jesus was now out in the open. Spurred by Jesus' popularity with the people, they began to plan for His murder. On His part, Jesus exposed the scribes and Pharisees publicly as pompous hypocrites. In a sweeping denunciation recorded in Matthew 23, He repeatedly called them "hypocrites," and described them as "blind guides," "blind men," "whitewashed tombs, which look beautiful on the outside but on the inside are full of dead men's bones and everything unclean," and as the murderers of the prophets. He denounced them: "You snakes! You brood of vipers! How will you escape being condemned to hell?" (Matt. 23:33, NIV). They were "blind fools" (Matt. 23:17, NIV) who destroyed the very law which they sought to uphold.

He concluded His denunciation with a lament over the city of Jerusalem, "O Jerusalem, Jerusalem, you who kill the prophets and stone those sent to you, how often I have longed to gather your children together, as a hen gathers her chicks under her wings, but you were not willing. Look, your house is left to you desolate. For I tell you, you will not see me again until you say, Blessed is he who comes in the name of the Lord" (Matt. 23:37-39, NIV).

Jesus Predicts the Destruction of Jerusalem

The disciples saw the drama of the triumphant entry and the open hostility of the Jewish leaders. They also saw how deeply Jesus loved the people and the city. But Jesus was more moved than anyone else. Knowing the destruction that would soon overtake the holy city, Jesus wept as He predicted: "If you, even you, had only known on this day what would bring you peace — but now it is hidden from your eyes. The days will come upon you when your enemies will build an embankment against you and encircle you and hem you in on every side. They will dash you to the ground, you and the children within your walls. They will not leave one stone on another, because you did not recognize the time of God's coming to you" (Luke 19:42-44, NIV).

Less than forty years later these prophetic words were fulfilled. After years of internal turmoil and conflict between Jew and Roman, the Roman army marched on Jerusalem, surrounded it, and starved it into submission. Then the Romans systematically looted and burned the city, slaughtering hundreds of thousands. Estimates indicate that as many as a million lives were lost during this period. In addition to destroying the city, the Romans pried loose the great stones of the beautiful temple, completed by Herod just six years before, and threw them into the valley to the southeast of Jerusalem. Today archeologists are still finding new evidence of this destruction. At this time (A.D. 70) a new period of wandering and persecution for the homeless Jew began. This has lasted until the twentieth century.

The Disciples' Questions

The disciples did not fully grasp what was happening nor what Jesus meant. They were bewildered by the fact that the Jewish leaders had rejected Jesus. How did this fit in with the Old Testament prophecies that

Jesus was to be their King and Savior? The whole incident of the triumphant entry into the city and the hostility of the Jewish leaders left the disciples full of questions.

After leaving Jerusalem, crossing the brook Kidron, and climbing the Mount of Olives to the east of Jerusalem, four of the disciples, Peter, James, John, and Andrew, came to Jesus privately and began to ask questions (Mark 13:3). They said, "Tell us, when will these things happen? And what will be the sign that they are all about to be fulfilled?" (Mark 13:4, NIV). Matthew's gospel indicates that they also asked questions about the end of the age, saying, "When will this happen, and what will be the sign of your coming and of the end of the age?" (Matt. 24:3, NIV). Actually, they were asking two important questions. First, when would Jerusalem be destroyed, and second, when would the present age end and the kingdom be brought in?

The disciples were attempting to reorganize their view of prophecy and were dimly beginning to understand that all prophecies regarding Jesus were not to be fulfilled at His first coming. Instead, it seemed rejection and suffering were ahead. If Jerusalem were about to be destroyed, how could the glorious period of the kingdom be brought in? Did this mean there would continue to be Gentile oppression of the Jews? How would it all really end?

Jesus' Answers to the Disciples' Questions

In studying the answers Jesus gave to the disciples as recorded in Matthew, Mark, and Luke, it becomes clear that He was talking about two specific periods in the future history of Jerusalem. One set of prophetic events would end in the destruction of Jerusalem, which we know occurred in A.D. 70 and fulfilled precisely the warnings and predictions contained in Luke 21:5-24. The second set of prophecies referred to the whole age following the destruction of Jerusalem. This age would have many miraculous signs which would culminate in the second coming of Christ to the earth (Luke 21:25-28).

Because the prophecies regarding the destruction of Jerusalem have already been fulfilled, it is quite easy for us to understand Jesus' predictions about it. Because of the rejection of the Jewish leaders, Jerusalem was facing destruction. Roman armies would surround the city, kill its inhabitants, burn the city, and destroy the temple, leaving the entire city in complete desolation. This would be the beginning of a renewed period of persecution and trial for the people of Israel. This period would have many important characteristics, which Jesus described in detail as recorded by Matthew and Mark.

The Present Age As Described by Jesus

In Matthew 24:4-14 Jesus gave nine characteristics of the age in general. It would include (1) many who claim to be Christ but are imposters, (2) "wars and rumors of wars," (3) famines, (4) "pestilences" (Matt. 24:7, AV), (5) earthquakes, (6) many martyrs, (7) many false prophets, (8) increasing wickedness and decreasing love for God, and (9) the Gospel of the kingdom to be preached to all nations. The last nineteen hundred years have demonstrated the accuracy of this prophetic analysis of the present age. Jesus knew what He was talking about and did not make any false predictions.

Exactly as Christ predicted, the present age has known much religious confusion and deception. War has followed war, and constant fear of war has plagued every generation. Great famines and epidemics have swept the world, killing millions. Probably more people have starved to death in the twentieth century than in any previous period. Religious persecution has marked all these years, especially persecution of the Jews. Families have been divided on the question of

85

their religious faith, and men in power have made countless attempts to stamp out religious beliefs with which they disagreed. The result has been literally millions of martyrs in the last nineteen hundred years.

The Accuracy of Jesus' Predictions

In the study of the detailed prophecies relating to the present age, we see an impressive consistency and accuracy in fulfillment of the prophecies. Anyone living today who attempted to prophesy the course of the next two thousand years of world history would face the utter impossibility of being accurate without a divine revelation. Yet Jesus was completely accurate, as history has demonstrated. But all the prophecies of Jesus have not yet been fulfilled. The end of the age is yet future. If history has recorded the literal fulfillment of Jesus' predictions to this point, can we project the same accuracy and literal fulfillment to the predictions that are yet future? This is the question which faces every intelligent person living today. Without these inspired predictions it is foolish to speculate. The only source of accurate information is the Bible itself. What is the future of the world? And what is the future of Jerusalem and the Jew? Jesus Himself gave detailed predictions and warned His hearers to watch Jerusalem.

7

Watch Jerusalem

Israeli soldier prays at the Wailing Wall, his rifle resting against the Wall.

UPI photo

On a Friday night at the beginning of the Jewish Sabbath, thousands gather at the Wailing Wall for the Sabbath prayers.

UPI

7

Watch Jerusalem

What Is the Future of Jerusalem?

A generation ago some people took the position that Israel was finished as a nation. Though individual Jews could be revived spiritually, they would never return to their land, never again occupy the city of Jerusalem, and never fulfill the glorious predictions of the prophets. These people believed that because the Jews had rejected their Messiah, God had rejected them.

Today this categorical dismissal of the future promises relating to Israel as a nation is no longer tenable. The facts are that Israel was returned to the land, was restored as a nation, and once again occupied her ancient capital, Jerusalem. But what is the future of Jerusalem? What is the order of events which will lead to the fulfillment of all the prophecies?

Many have pointed out that while Israel has been restored in part as a nation, the predictions of a glorious kingdom of peace and righteousness on earth have certainly not been fulfilled. The Jews have not recognized Jesus as their Messiah, and they certainly have not achieved peace, safety, and economic prosperity. While some are still disposed to write off the dramatic events in Israel's recent history as unrelated to the Bible, and others have been too eager to claim fulfillment of prophecy in the current situation, just what are the facts? What is the future of Jerusalem and the future of Israel?

The War of 1967

In the struggle of Israel after her founding as a nation in 1948, probably no more dramatic moment occurred than the sensational Six Day War of 1967. For the first time since A.D. 70 the ancient city of Jerusalem fell into Jewish hands. Jewish territory was greatly extended, including the portion of Jordan lying west of the Jordan River. What was the significance of this impressive advance for the state of Israel?

Many turned to the significant prophecy of Jesus in Luke 21:24, where Jesus was describing the course of the present age. Jesus said, "Jerusalem will be trampled on by the Gentiles until the time of the Gentiles are fulfilled" (Luke 21:24, NIV). With Israel once again in possession of her ancient city, was the time of the Gentiles now fulfilled?

Events which have occurred since 1967 have made it clear that Israel's present possession of Jerusalem is temporary. The hour of the glorious kingdom has not yet arrived. If it were not for the backing of the United States, Israel would soon be engulfed by the Arab world that surrounds it. This became especially evident in the war of 1973 when Israel survived only because of the prompt resupply of the weapons of war by the United States. Many other Scriptures indicate that Israel's final hour of triumph cannot come until Jesus Christ Himself returns from heaven in glory to deliver His people and to establish His kingdom of righteousness and peace on earth.

But what does the Bible teach concerning the future of Jerusalem and Israel, and how does this relate to tremendous prophecies yet to be fulfilled concerning a final struggle at the end of the age which will involve the whole world? Many of the answers to these questions are found in the prophecies of the prophet Daniel in the Old Testament. Human history is destined to have a dramatic climax which can only be understood by studying these prophecies and understanding their meaning for events which face the world today.

Daniel's Prophetic Outline of World History

The prophet Daniel was a young lad in his early teens when Nebuchadnezzar, the triumphant general of the Babylonian army, swept into Jerusalem in the summer of 605 B.C. In the course of the campaign, Nebuchadnezzar's father died, and Nebuchadnezzar became the young king of a tremendous empire. One of his first acts was to collect some of the choice young men in Jerusalem and carry them off to his capital in Babylon, partially as hostages and partially to serve him in various capacities. Among them was Daniel who, after several years of training in the school of Babylon, was prepared for a major role as an administrator in Nebuchadnezzar's empire.

One day Nebuchadnezzar had a horrible nightmare. In a dream he saw a great image standing near his bed and towering above him. The image had a head of gold; the upper parts of its body were silver; the lower part of its body was brass; the legs were made of iron, and the feet were of iron mixed with clay. Nebuchadnezzar was terrified by the dream and was sure it had some meaning.

Hastily his wise men were summoned. Whether Nebuchadnezzar intended to test them to see if their interpretation was genuine or whether he forgot the dream, he nevertheless demanded that the wise men tell him the dream and its interpretation. When they were unable to do so, in a rage he ordered their execution. Unfortunately, Daniel, who was not present, was classified as a wise man and was subject to the death order.

When word of this dramatic event reached Daniel and his three companions, Daniel requested time and promised to interpret the dream. Subsequently, God revealed Nebuchadnezzar's dream and its meaning to Daniel. The great image represented four great world empires, beginning with Babylon, Nebuchadnezzar's kingdom, represented by the head of gold. It would be followed by a second kingdom represented by the silver. The lower part of the body of brass represented a third kingdom. The legs of iron and the feet of iron and clay were a fourth kingdom. In his vision Nebuchadnezzar had seen a stone hit the feet of the image and destroy the whole image. Then the wind blew away the debris. Daniel indicated that ultimately there would be a fifth empire which would destroy all that was before it and fill the whole earth. In gratitude for Daniel's explanation of his vision, Nebuchadnezzar elevated Daniel and his three friends to important political positions.

About forty years later, after Nebuchadnazzar had died, Daniel himself had a vision recorded in Daniel 7. Here, the same prophetic outline was indicated by four great beasts. Babylon was described as a lion; the second empire of Medo-Persia was pictured as a great bear; the third empire, that of Alexander the Great, was described as a leopard with four heads and four wings; and the fourth beast was described as the most powerful of all with iron teeth and tremendous strength. Daniel the prophet himself named the first three empires and described in detail the course of the prophetic fulfillment. Of special interest, however, is the fourth empire and its end, presented in Daniel 7:7-28.

Just as Daniel predicted, Babylon was followed by Medo-Persia, and Medo-Persia was followed by Alexander the Great and the Grecian Empire. History has also recorded the rise and fall of the Roman Empire. These events are noted in the chart of prophetic events at the close of this chapter.

Two great puzzles are left unresolved in Daniel's

prophecy, however. First, the precise prophecies concerning the end of the fourth empire were never fulfilled in history. Second, Daniel predicted a fifth empire which would succeed the Roman Empire, a kingdom which would come from heaven. This, likewise, has never been fulfilled. If the predictions of Daniel were so accurately fulfilled up to the present time, should not one conclude that these prophecies will also have their fulfillment, perhaps in the near future?

The Gap in Daniel's Prophetic Vision

Like many other prophecies in the Old Testament, Daniel's prophecy of future events leaves a wide gap between the fulfillment until the time of Christ and the events which will end the age. It seems as if the prophetic fulfillment was, for all practical purposes, stopped with the first coming of Christ, and unfulfilled prophecy only deals with the events immediately preceding Christ's second coming. Because the Old Testament prophets did not deal in detail with the progress of the present age, Christ Himself in His prophetic ministry filled in the gap in Matthew 13 and again in Matthew 24:3-14. In these prophecies He made it clear that after most of the present age had run its course, then the great events of which the Old Testament prophets spoke would be fulfilled. It was this puzzling gap in Old Testament predictions that prompted the disciples' questions and Jesus' answers.

The description which Jesus gave of the present age in Matthew 24:3-14 has already been outlined in the preceding chapter. But what did Jesus have to say about the end of the age? How does this fit into the description of the climax of the age as given in the book of Revelation?

Confirmation of Daniel's Prophecy by the Apostle John

The book of Revelation from chapter 4 through chapter 18 reveals in detail the shattering events which will end the age. In general, it describes a period of unprecedented trouble, culminating in a great world war.

Most important, however, is the fact that the Apostle John, writing six hundred years after Daniel, confirmed and gave added detail concerning the end of the age as Daniel saw it. In Revelation 13 the same great world empire is described, the same world ruler, and the same climax — the coming of the kingdom from heaven at the time of the second coming of Jesus Christ. Like Daniel and many Old Testament prophets such as Zechariah, the Apostle John pictured Jerusalem as the center of the maelstrom which would draw all nations to the Holy Land for the final world conflict. It was Jesus, however, who defined the central role of Jerusalem in detail.

The Coming Desecration of the Holy City

The disciples had come to Jesus asking Him concerning signs of the end of the age and of the inauguration of His kingdom (Matt. 24:3). After giving them the general signs of the whole period in Matthew 24:1-14, Jesus revealed that Jerusalem itself would be the scene of the specific sign that the coming of the Lord was near. Jesus said, "So when you see standing in the holy place the abomination that causes desolation, spoken of through the prophet Daniel — let the reader understand — then let those who are in Judea flee to the mountains" (Matt. 24:15, 16, NIV).

What was this "abomination that causes desolation" which, according to Jesus, was mentioned by the prophet Daniel? This subject was introduced by Daniel in three different passages in his book. He prophesied, "And forces from him will arise, desecrate the sanctuary fortress, and do away with the regular sacrifice. And they will set up the abomination of desolation" (Dan. 11:31, NASB). Many interpreters find this prediction fulfilled about 350 years after Daniel's death by

94

a ruler of Syria known as Antiochus Epiphanes who ruled from 175 to 164 B.C. He was a fearful persecutor of the Jews and attempted to stamp out their religion. In the process, he desecrated their temple by offering a sow upon the altar and setting up a statue of a Greek god in the holy place. He killed tens of thousands of Jews who rebelled against him because of it. This event is referred to as "the abomination of desolation" because his acts were an abomination to a holy place such as the temple, and it desecrated or desolated the temple. With the help of this fulfillment of Daniel 11:31, it is possible to understand two other references of Daniel to a similar event which will take place in the future.

In Daniel 9:27 the future world ruler is pictured as putting "a stop to sacrifice and grain offering," and it is predicted that "on the wing of abominations will come one who makes desolate, even until a complete destruction" (NASB).

Again Daniel 12:11 predicts, "And from the time that the regular sacrifice is abolished, and the abomination of desolation is set up, there will be 1290 days" (NASB).

A careful study of these passages indicates that approximately three-and-a-half years before the second coming of Christ, the dictator in the Mediterranean will desecrate a future Jewish temple and stop the sacrificial worship of God being carried on in this temple. The events, therefore, will be a specific sign that the end is near and that the time of trouble for Israel and the world will immediately follow. Because this event will occur in Jerusalem, it again plainly shows that we should watch Jerusalem as the key to prophetic events of the end time.

This prediction is most significant, as it anticipates a rebuilding of the Jewish temple in Jerusalem by orthodox Jews and the renewal of ancient forms of worship prescribed in the Law of Moses. By his act of desecration, the ruler of the Middle East, who it seems, will have just assumed the role of being a world ruler, will dramatize his take-over of the world religiously as well as politically. The time that follows will be "the time of Jacob's trouble" (Jer. 30: 7, AV), a time of worldwide distress and persecution of the people of Israel, and will serve as the warning of impending catastrophes about to sweep the entire world.

In the action of stopping Jewish worship, the new world ruler will reveal himself as Satan's substitute for Jesus Christ, sometimes called Antichrist.

The Prophetic Pulse of Jerusalem

The prophecies about Jerusalem make it clear that the holy city will be in the center of world events in the end time. Most of the events just over the horizon were anticipated by Jesus' prophecy. In predicting these events, Jesus confirmed the accuracy of Daniel's predictions, both relating to Jerusalem and to the Gentiles. In our present rapidly moving world scene with the Middle East once again becoming the center of the stage, it becomes dramatically clear that these events may not be too far distant.

The world today may be in the last years of the times of the Gentiles. As international power shifts to Europe and to Mediterranean nations, the time is ripe for a new economic and political struggle and for new alliances and new leaders to emerge. Meanwhile, the conflict between Israel and the Palestinian Arabs will focus more and more attention on Jerusalem. Religious passions will dominate the situation, making difficult any realistic peace settlement. The eventual result may well be a forced settlement imposed upon the Jewish-Arab world which will, in effect, make Jerusalem an open and international city.

As signs that we may be moving into this period multiply, the direction of present world events also points to the conclusion that the coming of Christ for

His church, promised in John 14, may occur any day. The fulfillment of God's present purpose to call out those who believe in Jesus Christ from both Jew and Gentile to form a body of believers in this present age will open the door for these final world events, which will include the emergence of the Mediterranean leader, the formation of a federation of Middle East states, and a forced peace upon the Israeli-Arab conflict.

In all of these situations Jerusalem is the city to watch, as the city of prophetic destiny prepares to act out her final role. The total world situation may be expected more and more to be cast into the mold which prophecy indicates. From many indications it seems that the stage and the actors are ready for the final drama, in which Jerusalem will be the key.

Chart of Prophetic Events in History
Beginning With the Babylonian Captivity

605 B.C.	Fall of Jerusalem: beginning of Babylonian Captivity and the times of the Gentiles (Jer. 25:9-11; Luke 21:24).
586 B.C.	Destruction of Jerusalem and Solomon's temple (Jer. 52:12-13).
539 B.C.	Fall of Babylon: beginning of second empire of Medo-Persia (Dan. 2:39; 5:1-31; 7:5).
538 B.C.	Second return of the Jews to the Holy Land (Ezra 1:1 - 2:70).
515 B.C.	Rebuilding of temple in Jerusalem (Ezra 3:1-13; 4:21-24; 5:1 - 6:18).
445 B.C.	Rebuilding of Jerusalem walls (Neh. 1:1 - 6:16).
445-396 B.C.	City of Jerusalem rebuilt (Dan. 9:25).
331 B.C.	Beginning of third empire of Greece (Dan. 2:39; 7:6; 8:1-22).

242 B.C.	Beginning of fourth empire of Rome (Dan. 2:40; 7:7).
63 B.C.	Romans conquer Jerusalem.
20 B.C.	Romans begin rebuilding of Jerusalem temple.
5-4 B.C.	Birth of Jesus Christ (Isa. 7:14; 9:6-7; Micah 5:2; Luke 2:4-11).
A.D. 30-33	Death and resurrection of Jesus Christ (Ps. 16:19; 22:1-21; Isa. 53:3-12; Matt. 16:21; 17:22-23; 20:18-19; 27:33-50; 28:1-10).
A.D. 64	Jerusalem temple completed.
A.D. 70	Destruction of Jerusalem and temple: worldwide scattering of the Jews (Deut. 28:63-68; Matt. 24:1-2; Luke 21:20).
A.D. 1897	Zionist movement begins: Israel seeks home in ancient land.
A.D. 1945	Rise of Russia and communism to power.
A.D. 1946	Beginning of world government: United Nations formed.
A.D. 1948	Israel established as a nation in the land: third return, first phase of Israel's final restoration.
A.D. 1948	Beginning of world church: formation of World Council of Churches.
A.D. 1956	Israel extends territory.
A.D. 1967	Israel regains territory from Jordan River to Suez Canal, including Jerusalem.
A.D. 1973	Israel extends territory into Egypt and Syria, and negotiates withdrawal from Suez.
A.D. 1975	Egypt reopens the Suez Canal.

8

The Rising Tide of World Religion

8

The Rising Tide of World Religion

The Role of Religion in the End Time

Until our present generation, Western civilization was quite content with its scientific progress and its rational and evolutionary explanation of the world. Today, however, things are different. The secure and rational world of the past is already beginning to collapse. On a scale never thought possible in a modern world, astrology, psychic phenomena, witchcraft, and the occult are permeating our culture. This is the same world that formerly denied the reality and importance of anything science could not measure and control. A new, mysterious world of superstition, fear, and demonic activity is attracting millions of people.

Such a remarkable change in world attitude does not surprise students of biblical prophecy. In prophecies of the end time there are many indications that there would be an increasing trend toward religious deception and conflict as the end of the age neared. Christ Himself warned of false prophets and religious charlatans who would rise to prominence in a gullible world looking for some religious answer to the world's problems.

Religion has always played a major role in the history of the world. Many of the wars of the past as well as dramatic historical events have stemmed from religious conflict. Biblical prophecy indicates that religion likewise will have a major role in the dramatic events which will conclude the present age.

Present Religious Tensions in the Middle East

Middle East tensions are rooted in centuries of religious tradition. Israel is motivated to possess her ancient land by religious convictions, as well as by national and political aspirations. The Arab hatred of the Jew is partly racial but also clearly related to the religious issues which surround the possession of the city of Jerusalem.

The conflict extends even to the former Jewish temple site, where the Dome of the Rock stands as a symbol of Muslim religious faith. This Muslim shrine is a major obstacle to the rebuilding of a new temple for Judaism on the ancient temple site. Possession and access to this site, including the ancient Wailing Wall, was considered an important and sacred accomplishment of the war of 1967. The struggle to negotiate a settlement for Jerusalem involves important religious considerations which will frustrate direct negotiations between Israel and the Arab bloc.

The Church's Claim to Jerusalem

Although Israel and the Arabs seem to be the principal contenders for possession of Jerusalem, it must not be overlooked that Jerusalem is considered a holy city by millions of professing Christians in the world. Few Western leaders can afford to ignore the destiny of Jerusalem because of the strong sentiments of adherents of Protestant, Catholic, and Eastern Orthodox churches. The Christian faith, the Muslim religion, and Judaism all have their geographic origin in the Middle East. These monotheistic religions are all related in their heritage to the Holy Land, the city of Jerusalem, and the promises God gave to Abraham. Any settlement of the Middle East crisis must satisfy

102

not only Jews and Arabs but also millions of Christians around the world. The spirit of the Crusaders is still alive today, and Christians will not stand on the sidelines and let the Arab and Israeli world determine the future of Jerusalem. Obviously, some kind of united effort on the part of major world religions is the only practical solution to the problems of the Middle East and Jerusalem.

Is a United World Religious Effort Possible?

Until the twentieth century the contending forces of Judaism, the Muslim faith, and the Christian church seemed basically irreconcilable. Each of these faiths was born in strenuous times and survived in spite of persecution and opposition. The history of the Christian church itself seems to indicate the impossibility of ever uniting the Middle East or the world in a common religious effort.

Christianity was born in a time of political chaos with little opportunity for formal organization. When the church finally gained political recognition, it rapidly began to organize, nominate bishops, and hold church councils. In a time of political disorganization, the church was able to organize and be a stabilizing force in the entire Middle East.

In the eleventh century, however, the possibility of a united church was shattered by a split between the East and the West, resulting in the formation of the Orthodox church in the East and the Roman Catholic church in the West.

A few centuries later a further division arose in a protest against the Roman Catholic church, and the Protestant church came into being. The Protestant church, in turn, multiplied into hundreds of denominations throughout the world. By the end of the nineteenth century, the possibility of a united Christian church seemed as impossible as any union of Christian, Muslim, and Jewish faiths.

The last fifty years have witnessed an amazing reversal of church history in a movement toward a world church within Protestant Christianity. Preliminary meetings held in 1925 and 1927 resulted in a temporary ecumenical council formed in 1938. Out of this came the World Council of Churches, which was organized at Amsterdam in 1948. In this new affiliation, many denominations and millions of Christians joined in an effort to form a super-church. Almost from the beginning, hope was expressed of bringing back under one church organization all branches of Christianity, including Protestant, Roman Catholic, and Eastern Orthodox churches. In the process, however, strict adherence to biblical doctrine and orthodox theology was sacrificed for organizational unity. Students of biblical prophecy noticed at once the remarkable similarity between the world church movement and biblical prophecy concerning religion in the end time.

The Prophetic Warning Against False Prophets

Shortly before he died as a martyr, the Apostle Peter, the leader of the twelve disciples, wrote, "But there were also false prophets among the people, just as there will be false teachers among you. They will secretly introduce destructive heresies, even denying the sovereign Lord who bought them — bringing swift destruction on themselves. Many will follow their shameful ways and will bring the way of truth into disrepute" (2 Pet. 2:1, 2, NIV).

As Peter made clear, the problem was not a minor dispute over some fine point in theology but involved the central doctrine of Jesus Christ — who He was and what He did when He died upon the cross. Peter went on to warn that these false prophets would be hypocrites parading in religious garb, when actually they are immoral, corrupt, and depraved in their moral stan-

104

dards. In his final word, he described their scoffing at the truth of Jesus' coming again (2 Pet. 3:3, 4).

In like manner, in the epistle written shortly before his martyrdom, the Apostle Paul warned Timothy, "Preach the Word; be prepared in season and out of season; correct, rebuke and encourage — with great patience and careful instruction. For the time will come when men will not put up with sound doctrine. Instead, to suit their own desires, they will gather around them a great number of teachers to say what their itching ears want to hear" (2 Tim. 4:2-4, NIV). Similarly, the entire epistle of Jude is dedicated to warning against false teachers.

Disputes in theology are not new, and departures from strict adherence to biblical truth have occurred repeatedly throughout the centuries. In the twentieth century, however, change in theology has been more rapid and more devastating to biblical faith than ever before. The common agreement of the church for centuries that the Bible is indeed the Word of God has been abandoned by many who claim to be leaders in the church. For the first time, theologians have proclaimed "God is dead," and atheism has been advanced as an alternative to biblical faith. In the religious field the stage is also set for fulfillment of these prophecies as the world moves to its final hour.

The Rise of the Occult

One of the significant prophecies of the Apostle Paul was addressed to Timothy: "The Spirit clearly says that in later times some will abandon the faith and follow deceiving spirits and things taught by demons" (1 Tim. 4:1, NIV). In this prophecy the apostle predicted not only that men would abandon the Christian faith in the end time but also that a prominent feature would be that men would follow spirit beings and the teachings of demons, messengers of Satan. These deceiving spirits would seek to seduce the world into accepting falsehood as the truth.

A generation ago it would have seemed inconceivable that any large number of people in a modern society would believe in spirits and demons, except for Christians who accepted the existence of spirit beings on the testimony of Scripture. What was unthinkable a generation ago is widely accepted today, even on the part of those who have no regard for the Bible.

The idea of psychic communication is now considered scientifically plausible, with experiments continuing to explore extrasensory perception. Fortune-tellers and mediums have always claimed to have a special kind of communication with the dead as well as the living. These mediums claim that spirits of those who have died can return as guides to give special insight into the past, present, and future.

The idea of spirits as helpers and guides has popularized the spiritism movement, built on the idea that communication with spirits can be a positive experience. More and more people have sought this experience through various rituals and seances, and new religious leaders, claiming to have supernatural power as mediums, have emerged and become prominent.

While much of this has occupied those who are outside the Christian church, the interest and involvement of men like the late Episcopal Bishop Pike have brought spiritism into the headlines. Newspapers and magazines now feature prophecies for each year, and psychics are becoming celebrities. Movies such as *The Exorcist*, which graphically show the power of demons, have attracted millions of viewers.

The entire trend toward experimentation with unseen spirits, often aided by drugs, is a frightening example of satanic deception. All these things are an amazing development following the pattern outlined in the prophetic Word for the time of the end. With Christian faith and morality being discarded by many, the new trend toward spiritism and the occult is an obvious satanic substitute for biblical Christianity.

What Happens Next?

The Apostle Paul warned of the coming time of the end in 2 Thessalonians 2:1-12. The Thessalonians had been confused by false prophetic teaching. Paul clarified the situation by describing four future prophetic events in a definite order which provides a calendar of events for the end time. For informed observers, these events signal the countdown to Armageddon.

First, there would be the rise of unbelief and departure from the faith within the church. Paul wrote, "Concerning the coming of our Lord Jesus Christ and our being gathered to him, we ask you, brothers, not to become easily unsettled or alarmed by some prophecy, report or letter supposed to have come from us, saying that the day of the Lord has already come. Don't let anyone deceive you in any way, for that day will not come until the rebellion occurs and the man of lawlessness is revealed, the man doomed to destruction" (2 Thess. 2:1-3, NIV). By "rebellion" Paul meant departure from the faith. Our English word "apostasy" comes from the Greek word in the New Testament translated "rebellion."

Second, the Holy Spirit, who now restrains evil in the world and who indwells the church, will be taken away at the time of the rapture. Obviously, the removal of every Christian in the world who is indwelt by the Spirit will release a flood tide of evil such as the world has never seen. It will allow the immediate take-over of the world church by those completely devoid of Christian faith and will allow other forces of evil a free hand in human history.

Third, at this point in the calendar of events, Satan's man of the hour will emerge. Paul referred to this man as "the lawless one." Paul wrote, "And now you know what is holding him back, so that he may be revealed at the proper time. For the secret power of lawlessness is already at work; but the one who now holds it back will continue to do so till he is taken out of the way. And then the lawless one will be revealed, whom the Lord Jesus will overthrow with the breath of his mouth and destroy by the splendor of his coming. The coming of the lawless one will be in accordance with the work of Satan displayed in all kinds of counterfeit miracles, signs and wonders, and in every sort of evil that deceives those who are perishing" (2 Thess. 2:6-10, NIV).

The emergence of this wicked character, designed by Satan to be a substitute for Jesus Christ, is the signal for the beginning of all types of satanic works that will deceive the entire world and eventually cause men to worship this person as their god.

Fourth, the period of evil introduced by this "lawless one" will climax in the second coming of Christ, when the world will be judged and evil will be destroyed. This event coincides with the battle of Armageddon and the judgments that will attend the second coming of Jesus Christ to the earth.

The Emergence of the Super-Church

Fear and panic will accompany the disappearance of true believers from the earth at the time of the rapture. This will amplify the world's desire for a strong religious organization to bring order from that religious chaos. With millions of people disappearing, natural disasters, a rapid rise in demonic power, and false prophets performing signs and wonders, the world will be grasping for something that is secure. In desperation, the masses will turn to the world church for help.

In the absence of the redeeming presence of any true believers, the Catholic, Protestant, and Orthodox churches will combine into a powerful religious and political institution. The super-church will be able to command the obedience and devotion of hundreds of millions throughout the world and will have power to

108

put to death those who resist its demands for adherence. The new world church will be in alliance with the political powers of the Middle East. This combined effort will prepare the way for a new government with absolute power over the entire world.

This unholy alliance is portrayed symbolically in Revelation 17, which describes a wicked harlot riding a scarlet-colored beast. For centuries, expositors have recognized the harlot as the symbol of religion and the scarlet beast as representing the political power of the Mediterranean Confederacy in the end time. While their alliance will bring a temporary stability to the world, it will also create a blasphemous religious system which will lead the world to new depths of immorality and departure from true faith in God.

One of the fruits of their combined power will be the forced peace with Israel to be discussed later. While there will be some show of religious tolerance in permitting the revival of Judaism, it will be short-lived and will continue only as long as the peace treaty is observed. In other parts of the world, persecution and martyrdom apparently will follow the extending power of the combined political and ecclesiastical alliance. But the final form of world religion will emerge later.

Communism As the Key to the Final World Religion

One of the phenomena of the twentieth century has been the rapid growth of communism which has swept more than one-third of the world into its clutches. Although the political idealism of communism does not seem to be perpetuated in the end time, the communist dedication to atheism, materialism, and military power will prepare the way for the final form of world religion.

After describing the alliance of the church and the political power of the Middle East during the period of the peace treaty with Israel, the Apostle John revealed that the harlot-church would be destroyed and burned with fire by the ten kings associated with the Mediterranean dictator. This apparently will come at the beginning of his bid for world power, approximately three-and-a-half years before the second coming of Christ. Scripture makes it quite clear that the final form of religion will be atheistic, materialistic, and man-centered. To some extent, it will be similar to the pattern of Roman emperors, who deified themselves as gods.

World communism today presents a similar atheistic, materialistic, and religious system, which is an obvious preparation for this final form of world religion. According to Daniel 11:36-38, the final world ruler will disregard all previous gods and will honor only "a god of fortresses," referring to military and materialistic power personified. The final form of world religion, empowered by Satan himself, will be the worship of the world ruler who is Satan's substitute for Jesus Christ. In the world today there is already a world church movement preparing the world for a world church. And the communist movement is preparing the world for the final form of atheistic religion. This makes it clear that the stage for the end time is already set.

Never Before

Never before in history have all the factors been present for the fulfillment of prophecy relating to end-time religious trends and events. Only in our generation have the combined revival of Israel, the formation of a world church, the increasing power of Muslim religion, the rise of the occult, and the worldwide spread of communism's atheistic philosophy been present as a dramatic setting for the final fulfillment of prophecy. As far as world religion is concerned, the road to Armageddon is already well prepared, and those who will travel to their doom may well be members of our present generation.

9

The Coming Middle East Peace

United Nations peacekeeping troops pass Israeli soldiers
evacuating from Suez City, Egypt.

9

The Coming Middle East Peace

The World Clamor for Peace

All signs point to the necessity of a coming peace settlement in the Middle East. The industrialized nations cannot continue to tolerate the disruption of oil supplies in the world market. The struggles between Israel and the Arab nations cannot continue to push the entire world repeatedly to the brink of nuclear war. The overwhelming weight of population advantage, proximity to Russia as supplier of arms, and the tremendous power of the oil blackmail give the Arab world the ultimate advantage at the peace table. In Israel, changing internal politics may demand some demobilization of the military and dependence on outside guarantees for Israel's security. The United States will be reluctant to give such guarantees which might cause another military involvement overseas. The international political leader who can give Israel these guarantees and force dissident Arab factions to accept a final peace will gain world recognition overnight.

Peace and the Final Countdown

A peace settlement in the Middle East is one of the most important events predicted for the end time. The signing of this peace treaty will start the final countdown leading to Armageddon and then introduce the new world leader who will be destined to become world dictator — the infamous Antichrist. According to Daniel 9:27 the last seven years leading up to the second coming of Christ will begin with just such a peace settlement. The same passage describes a covenant to be made between the nation of Israel and the prince who will rise to power (Dan. 9:26). While the details of the covenant are not given, it will be an attempt to settle the Arab-Israeli controversy which has focused world attention on the Middle East. It may well take the form of a forced peace settlement in which Israel returns much of the land conquered through war in exchange for strong international guarantees for Israel's safety and prosperity.

The Middle East Market

Scriptures describe a world situation, not unlike what we have today, preceding such a peace settlement. First there will be an alignment of ten nations in the Mediterranean. In Daniel 7:7, 24 this is symbolized by ten horns on a beast that represents the last world empire, the Roman Empire in its final and revised form. Many interpreters of biblical prophecy have felt that the Common Market of Europe will be a fulfillment of this predicted alignment of ten nations. Whether the Common Market is the preliminary form of this ten-nation group or whether it is the forerunner of another confederacy of nations is difficult to predict. The final confederacy of ten nations must have the economic and political power necessary to control the Mediterranean. The final leader to emerge must eventually be able to seize control of three of these nations in a direct take-over.

The current economic situation precipitated by the energy crisis may soon create a new, more truly Mediterranean Confederacy, which will include some Arab and European countries in an alliance similar to the Common Market. In the final analysis, this will be a

114

confederation including both economic interdependence and military alliance. The trends of international intrigue in Europe and the Middle East clearly point to the necessity of an alliance of nations to bring peace and economic prosperity to that area. Changes necessary to make the present situation conform to the prophetic anticipation of Daniel could take place rapidly in the present world scene.

Israel Ready to Negotiate

Certain conditions, however, are absolutely necessary before the final peace settlement can occur. First, Israel has to be back in the land before such a peace treaty would even be possible. Israel's return to the land in 1948 and her recognition as a political state constitute an important first step. The military conflict and constant turmoil ever since have set the stage for just such a peace treaty. Without great world pressure for Israel to accept a compromise peace, any type of forced peace settlement would also seem unthinkable. The world energy crisis, accompanied by the new independence and wealth of the Arab states, has created a compelling movement toward a final peace covenant. The fear of an international economic disaster will cause the more sensible Arab countries, Israel, and the industrialized countries of the West to seek a final agreement at any cost. These realities are now present in our world.

A New World Leader to Emerge

A second set of realities predicted in Scripture concern a leader who will arise in the Middle East. His emergence may well come as an event immediately preceding the signing of a peace settlement. The Middle East today is looking for just such a world leader, one who will not only captivate the Middle East and gain its concurrence in his plan and program, but one who will ultimately rule the world. Henry Kissinger's attempt to begin peace discussions were only helpful

Israeli Defense Minister Moshe Dayan with U.S. Secretary of State Henry Kissinger just prior to the Geneva Middle East peace conference in December 1973.

in stopping the fighting. The new focus of attention will soon be on the struggle for key leadership to emerge in Europe and the Arab countries. The future man of the hour is awaiting the proper moment to make his move. As such a peace settlement seems to be imminent in the world, so the emergence of this leader also can be expected momentarily. Before the emergence of this world leader, the promise of Christ to come for His church will be fulfilled. International events in the Middle East thus make the rapture of the church a matter of immediate expectation.

What form will the coming peace settlement take? In view of the many surprises in the Middle East, it is hazardous to prophesy the precise form of such a final peace settlement. Whatever its preliminary form, ultimately it must give Israel security from attack and freedom from the constant state of military defense. It is very possible that an international peacekeeping force and secure boundaries may be guaranteed by the new and powerful Ten-Nation Confederacy. A general disarmament in the area may also be part of the agreement.

The key issue in negotiations will be the city of Jerusalem itself, which Israel prizes more than any other possession. Undoubtedly, there will be a strong attempt to make Jerusalem an international city with free access not only for Jews but for Christians and Muslims as well. The temple area may be internationalized, and Israel's territorial conquests will be greatly reduced. In the light of Arab power and the isolation of the United States as the sole supporting force behind Israel's continuity as a nation, it seems that any settlement short of this will not satisfy the Arab world.

How soon will such a peace settlement come? No one can hazard a guess. But come it must; here the Scriptures are clear. There will be a covenant between the Mediterranean leader and Israel. It will be a covenant which will permit Israel to continue and to renew her religious ceremonies, including the building of a Jewish temple and the reactivation of Jewish sacrifices. All of this was anticipated in the prophecies of Daniel 9:27 and 12:11 and was implied in the prophecy of Christ Himself relating to the stopping of the sacrifices when the covenant is broken (Matt. 24:15).

Made to Be Broken

Rumors from Arab conferences have indicated that any form of peace settlement now would only be a temporary position of advantage in preparation for a possible later war with Israel. This tactic has been necessary to appease radical Palestinian factions who oppose any permanent settlement. Unfortunately the final peace will be much the same — it will be a peace made to be broken. The peace settlement prophesied in Scripture will be a bid for world recognition by the new leader of the Mediterranean Confederacy. The peace will be observed only for three-and-a-half years and will be a major stepping-stone in his rise to world power. The peace settlement will be destined to be broken, bringing devastating consequences for the world and terrible persecution for Israel. This breaking of the covenant will mark what Christ called the "great tribulation" (Matt. 24:21, AV), a time which Jeremiah referred to as "the time of Jacob's trouble" (Jer. 30:7, AV).

Waiting for the Prince of Peace

The final persecution of Jews during the time of Jacob's trouble will awaken Israel's understanding to what has taken place. All the hopes and illusions of the past will be stripped away. This clear fulfillment of prophecy will lead to the startling realization that the first coming of the Messiah is past and that His second coming is near. In the horror of the last three-and-a-half years of great tribulation, these new believers will cling to the hope of Christ's second coming. Jesus prophesied that He would not come again until Israel would say, "Blessed is he who comes in the name of the Lord" (Matt. 23:39, NIV). And so it will be that no permanent peace can ever come until the Prince of Peace, the Lord Jesus Christ, returns in power and glory to reign over the earth.

10

The Threat of Russian Intervention

The Middle East area. Notice the position of Russia and especially Moscow in relationship to Israel.

10

The Threat of Russian Intervention

The Fear of Russian Invasion

One of the major factors in the Middle East conflict has been the threat of Russian intervention. Just as the power and military weapons of the United States have assured Israel's survival up to the present time, so the power of Russia and her military might are the shadows cast over the Middle East situation. It was this threat that prevented the exploitation of Israel's success in 1956. It was this fear which caused the cease-fire in 1973 and prevented Israel from capitalizing upon her strategic position. The cease-fire came at a time when the Egyptian Third Army was almost surrounded. The isolation of the Third Army would have allowed the Israeli army and air force to advance deep into Egypt and eventually Syria. The limiting factor in each of these wars has been the fear that overwhelming Israeli military success would bring a Russian invasion, creating a direct confrontation between Russia and the United States and the possibility of nuclear war.

The fear of a Russian invasion has never been entirely unfounded. Russia has been deeply committed to the support of the Arab countries and would not have waited on the sidelines if one of the major countries was ready to collapse. The presence of the powerful Russian fleet plus the close proximity of Russia to the area would make a direct military intervention possible. Prophecy, in fact, has predicted a time when this will actually take place. Ezekiel described it in these words: "Therefore, prophesy, son of Man, and say to Gog, Thus says the Lord GOD, On that day when My people Israel are living securely, will you not know it? And you will come from your place out of the remote parts of the north, you and many peoples with you, all of them riding on horses, a great assembly and a mighty army; and you will come up against My people Israel like a cloud to cover the land. It will come about in the last days that I shall bring you against My land, in order that the nations may know Me when I shall be sanctified through you before their eyes, O Gog" (Ezek. 38:14-16, NASB).

The Declining Power of the Ruble

Russia has used various types of financial and military aid to establish her influence in the Middle East but her role has been tenuous at best. Technically ordered out of Egypt in 1972, Russia again became fully involved in 1973 and 1974 as a result of her military support of Syria and Egypt. Russia has poured approximately ten billion dollars of military hardware into the Middle East during the last ten years and has depended on extensive financial aid to achieve her goals.

With new realities in the Middle East, the power of the Russian ruble will not be sufficient to keep the Arab-Soviet friendship alive. The Arab bloc has begun to use its new financial strength to negotiate long-term agreements, exchanging oil for military hardware from Western European countries. As the Arab financial independence grows, Arab countries will be able to purchase sophisticated weapons from almost any country in the world. Agreements like this have already been made with France and are being negotiated with other countries. Iran and key Arab countries will

122

soon have some of the best-equipped armed forces in the world. The Arabs' dependence on Russia for military weapons and technology will be a thing of the past.

The Russian Diplomatic Gamble

The success or failure of Russia's Middle East policy will determine her future as a major world power. Russia's involvement in the Arab-Israeli conflict was obviously not out of singular concern for the welfare of the Arab world. Russia has depended heavily on her naval power to maintain her military and commercial presence throughout the world. Successful manipulation of the Arab countries has been necessary to continue Russia's freedom of movement in the Mediterranean and hopefully, through the Suez Canal, in the Indian Ocean as well. If Russia is to continue as a world power, her Baltic fleets must have freedom of movement west into the Atlantic and east, through Suez, into Southern Asia.

Since the countries of the Middle East are Russia's near neighbors, some control of the area is essential to Russian national security. The economic necessity of a continued supply of crude oil for the Soviet Bloc is also a part of the strategic value of involvement in the Middle East. As Russia's current strategy of diplomatic and financial involvement in the area begins to fail, the threat of a direct military move to control Middle East policy will undoubtedly increase. With the United States in no position to challenge Russia halfway around the globe, and with a spirit of isolation restricting the possibility of United States military involvement in distant lands, Russia may soon be in a position to risk a power play. Russia has a close proximity to the Middle East and will soon be prepared to gamble for the high stakes involved in the control of this area of the world.

From the standpoint of the prophetic Word, it is clear that Russia has no more real and vital union with the Arab world than the Arab world has with the United States. Russia is obviously in the conflict to maintain her own interests as a world power, and the Arab world is exploiting this friendship for its own selfish purposes. Prophecy has predicted that the Arabs and Russians will ultimately come to a parting of the ways.

The world energy crisis has introduced an important element to the alignment of nations in the Middle East. Both Russia and the United States have oil production and energy resources of their own, but European nations must depend entirely on Arab oil. It seems reasonable to assume that this will ultimately cause European nations to seek a more active role in the Middle East. European nations will be willing to make industrial and political concessions that Russia and the United States simply will not be willing to make. In the end, both countries may become isolated from the new power alignments forming between the Middle East and Europe.

Prophecy gives no indication of an active role for either the United States or Russia in the final peace settlement to be made in the area. During end-time events Russia will be excluded from participation in Middle East affairs and finally will resort to war in an attempt to gain control of the area. It is this invasion, described in Ezekiel 38 and 39, that will disrupt the Mediterranean Confederacy's peace covenant, swing the balance of world power sharply in favor of the Mediterranean leader, and allow him to declare himself world dictator.

Russia's Coming Invasion of Israel

The prophet Ezekiel described the time when Russia will make a military bid for power in the Middle

East. The prophetic chapters of Ezekiel 38 and 39 present many problems to the interpreter, but they clearly describe a horde of armed might invading Israel from the north. The names given to the leader, the country, and the cities, as well as the clear description of armies out of "the remote parts of the north" (Ezek. 38:15, NASB), could only refer to what we know today as Russia. The war is strange, as there is no record of an opposing army. The opposition is not from the Israeli army. The land is described as "restored from the sword" (Ezek. 38:8, NASB), enjoying a time of peace and prosperity in Israel with the inhabitants "living securely, all of them" (Ezek. 38:8, NASB). Ezekiel intended his prophecy to be a description of the "last days" (Ezek. 38:16, NASB) with Israel regathered, enjoying what it believes to be a permanent peace. Part of the motive for the Russian invasion is that Israel will be an easy spoil. This will be Russia's bid for power against the Mediterranean confederacy of nations that have guaranteed Israel's security.

The invading army is described as "a great assembly and a mighty army" (Ezek. 38:15, NASB). The Bible describes the use of primitive weapons — horses and horsemen clothed in full armor. Some believe these were the best words available to the prophet to describe modern technological warfare depending largely on tanks and armor, but a future disarmament combined with the energy crisis and the proximity of Russia to Israel would make a mounted cavalry assault especially useful for a sneak attack. Russia is one of the few world powers that continues to maintain a large cavalry. The Russian army in the invasion will be joined by other smaller countries also isolated from the prosperity and industrial growth of the countries within the Mediterranean Confederacy.

Russia's Downfall

Whatever the solution to these interpretive problems, the decisive war and the outcome of the battle are crystal clear. The invading army, led by Russia, will descend on Israel during a time of peace and prosperity. The invaders will be completely destroyed by a series of catastrophes outlined in Ezekiel 38:18-23. The invaders will be destroyed by a great shaking of the earth, which will be combined with an overflowing rain of hailstones, fire, and brimstone. Whatever this means, the army will be destroyed, and seven months will be occupied just in burying the dead.

God will intervene directly in history to save Israel. Many in Israel and throughout the world will see this event as an act of God. For some it will mark the beginning of a true belief in God, persecution, and martyrdom. For the rest of the world it will be the beginning of a torrent of catastrophic events, destroying the balance of power in the world and allowing the Mediterranean leader to declare himself world dictator without immediate opposition. It is very possible that the Mediterranean leader, the Antichrist, will not only step into the power vacuum created by this event but will also claim that he, in fact, destroyed the invading army.

Is the Russian Invasion Near?

One of the important questions in the interpretation of Ezekiel 38 and 39 is where to place the invasion in the sequence of events that lead up to Armageddon. Scholars have disagreed, some placing it as an event that could occur at any time, others making it a part of the battle of Armageddon itself. A few interpreters place it at the beginning of the millennial kingdom as an act of rebellion against Christ, and still others relate it to the rebellion at the end of the millennium recorded in Revelation 20:7-9.

According to Ezekiel, the invasion will come in the last days when Israel is enjoying a time of peace and prosperity. Israel is described as unarmed — a land of unwalled villages lulled into a sense of security. This has certainly not been true of Israel since its establishment in 1948. Since that time the tiny nation has been in a constant state of alert.

When will peace come? The predicted period of peace will come when the Mediterranean leader emerges to sign a peace covenant and settle the current Middle East strife. The emergence of this leader and his future is the subject of the next chapter. The most convincing arguments for the time of the Russian invasion place it during this covenant to be negotiated between the coming Mediterranean ruler and the nation Israel.

All other interpretations of Ezekiel 38 and 39 encounter serious problems. The invasion is unlikely to occur immediately following the second coming of Christ when the armies of the world have already been destroyed at the battle of Armageddon. The particular invasion described in Ezekiel 38 and 39 does not seem to fit the end of the millennial reign of Christ, even though Revelation 20:8 also mentions "Gog and Magog." The whole world, rather than only Russia, will be involved in this final rebellion against God at the end of Christ's thousand-year reign on earth. Immediately after this rebellion, the present heaven and earth will be destroyed and the eternal state will begin. By contrast, in Ezekiel 38 and 39 life has continuity. The passage includes details about the burying of the dead for seven months and the period of years before the area is free from the wounds of the disaster.

The details Ezekiel gave for the invasion and the events which will follow also do not fit the period at the end of the great tribulation just before the battle of Armageddon. The last part of the tribulation will be a time of unprecedented disaster, persecution of Jews and believers, and horrifying turmoil in the world. This will not be a time when Israel is at rest, secure in unwalled villages, prosperous with gold, silver, cattle, and goods. Taking these factors into consideration, it seems clear that the battle will come when Israel has been lulled into the false security of the peace agreement signed by the leader of the Mediterranean Confederacy. This peace treaty will be signed by the leader representing the combined economic and military power of the ten nations which will finally make up the Mediterranean Confederacy. With these international guarantees, Israel will turn her energies toward increased prosperity rather than defense — only to have the peace covenant shattered in less than four years.

Russia's Loss — The Antichrist's Gain

As will be seen from the next chapter, Russia's attack on Israel will not simply be an attack on Israel alone, but it will be a challenge to the peace and protection promised by the new Mediterranean leader. The emergence of the powerful new alignment of Mediterranean nations will by this time have seriously limited Russia's role in the world. The attack by Russia will be a desperate attempt to recoup her position as a world power with influence over the Middle East. The invasion of Israel will be enacted as a direct confrontation between Russia and the new Mediterranean leader. With the United States in isolation, the balance of power in the world will have shifted as a balance between the Mediterranean leader and his allies on one side and Russia and her allies on the other. Because of this, the destruction of the Russian army becomes tremendously significant. The balance of power will then fall dramatically into the hands of the Mediterranean leader. This greatly changed world situation will make him the most powerful leader in the world, leading to important and earth-shaking events which will soon follow.

The threat of Russian intervention in the Middle East is real. In the immediate future her financial diplomacy in the Middle East will not be sufficient to give her a major role in determining the area's future. Her eventual bid for power will be a military disaster.

In the end, Russia will reap a terrible judgment from God for her atheism, blasphemy, and persecution of Israel. Like many nations before, Russia will feel the curse of God promised upon all those who curse His chosen people.

RUSSIA'S INVASION OF ISRAEL
AS DESCRIBED IN EZEKIEL 38-39

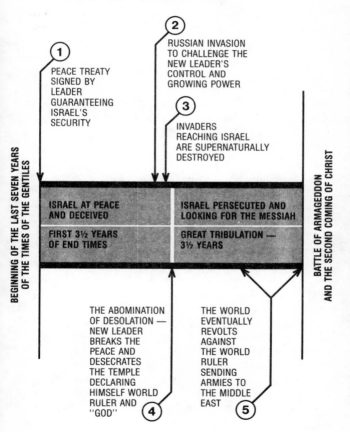

(1) PEACE TREATY SIGNED BY LEADER GUARANTEEING ISRAEL'S SECURITY

(2) RUSSIAN INVASION TO CHALLENGE THE NEW LEADER'S CONTROL AND GROWING POWER

(3) INVADERS REACHING ISRAEL ARE SUPERNATURALLY DESTROYED

BEGINNING OF THE LAST SEVEN YEARS OF THE TIMES OF THE GENTILES

ISRAEL AT PEACE AND DECEIVED

FIRST 3½ YEARS OF END TIMES

ISRAEL PERSECUTED AND LOOKING FOR THE MESSIAH

GREAT TRIBULATION — 3½ YEARS

BATTLE OF ARMAGEDDON AND THE SECOND COMING OF CHRIST

(4) THE ABOMINATION OF DESOLATION — NEW LEADER BREAKS THE PEACE AND DESECRATES THE TEMPLE DECLARING HIMSELF WORLD RULER AND "GOD"

(5) THE WORLD EVENTUALLY REVOLTS AGAINST THE WORLD RULER SENDING ARMIES TO THE MIDDLE EAST

129

11

The Coming World Dictator

11

The Coming World Dictator

The Dream of Conquerors

In the history of the world many a great man has dreamed of someday having the world at his feet. In the Middle East many great nations have risen and fallen. Egypt was a powerful nation in the Middle East fifteen hundred years before Christ. Then came Assyria with its capital at Nineveh. Nebuchadnezzar, the great king of Babylon, carved out a great empire which included most of western Asia. About three hundred years before Christ, Alexander the Great swept with his armies as far east as India and for a brief time ruled all of western Asia, northern Africa, and southeast Europe. At the time of Christ the Roman Caesars had subdued southern Europe, northern Africa, and western Asia to form the greatest empire ever known in history. In more modern times Napoleon dreamed of world conquest only to have his dreams shattered at Waterloo. In the twentieth century Hitler dreamed of someday having the world at his feet.

But none of these great men of the past ever really conquered the entire world. The military conquerors of the ancient world at best conquered portions of Europe, North Africa, and western Asia, but no world ruler has ever controlled all of the three major continents related to the Middle East, much less the nations of the western hemisphere — North America,

Central America, and South America. Until our generation the world was not really ready for a universal government.

A Shrinking World

The advances of modern technology in communication, transportation, and the weapons of war have suddenly shrunk our world. A missile can reach any part of the world in less than thirty minutes. Television and radio provide instant communication. A world in which men could once live in a measure of isolation is now geographically one. Every major event, whether it is an oil embargo, the development of a new nuclear weapon, or the threat of starvation, creates waves on the sea of mankind and pounds on distant shores. Men and nations can no longer live in isolation.

In World War I for the first time steps were taken to form an international body to prevent further world wars. The resulting League of Nations, however, was unsuccessful. With Congress already disillusioned by the practical results of World War I, the United States was not ready for such a role in world affairs. The idea of world government seemed unnecessary for a country oceans away from problems in Europe and Asia. But the earth was growing smaller, and America's isolation could not last for long. In 1941 the United States once again was drawn into a world war. The attack on Pearl Harbor marked the end of America's attempt to ignore her responsibilities in a shrinking world.

The Advent of the Bomb

By the end of World War II, a frightening new dimension had been added to the need for a world government. While the war itself was devastating and extended throughout more of the world than any previous war had, it ended with the atomic holocaust of Hiroshima and Nagasaki. With the advent of the atomic bomb, no nation or people could ever be safe again. The destructive capabilities of war had increased to a degree never before considered possible. Nuclear destruction threatened not only military forces but entire cities and civilian populations. Now a war of days or weeks could destroy the entire world. The time had come for a new international world organization.

World Government — Man's Last Hope

Out of the desire to avoid a major conflict involving atomic weaponry, the United Nations was born in 1946. While at first it included only a portion of the nations of the world, the United Nations soon grew to include most of them. With the admission of Red China, all the major nations and most of the smaller nations became part of this international structure.

But the United Nations lacks the power to prevent war. During its history its weakness has been demonstrated again and again. Its failure to solve the conflict in Southeast Asia and its continued inability to prevent flare-ups in the Middle East have demonstrated that the United Nations is not the final answer.

Still, some form of universal government seems the only hope for a world that can easily destroy itself. The problems facing the world — nuclear war, overpopulation, starvation, pollution, and economic instability — are international problems. Even small countries can now upset the important web of international interdependence, as demonstrated by the energy crisis in 1973. New economic problems face the world. The international distribution of resources and the control of currencies must necessarily be regulated by some type of world agency. The United Nations, the Common Market, and the World Bank are only the beginning of a quest for some solution to the world's increasing problems.

Many international leaders and intellectuals believe that a strong and effective world government is the only hope for the survival of man on this planet. As these attitudes are increasingly expressed in our time, it is important to ask what the Bible has to say on this subject. As a matter of fact, the prophets anticipated just such a state of affairs in the world as the end time approaches.

Daniel's Prophecy of World Empires

The prophets may never have realized the modern reasons for a world government, but they did predict that history would end in one central government which would embrace the entire world. The Bible not only predicted the rise and fall of the important world empires that have passed but with prophetic accuracy has described the events that will lead to a final world government before the second coming of Christ. The prophet Daniel, for instance, described the series of developments that will bring the entire earth to a final world government, which will reach its climax when Christ returns to judge the nations and set up His kingdom on earth (Dan. 7).

Daniel predicted with accuracy the progression of preceding world empires. The first empire described by Daniel was that of Babylon, which conquered Jerusalem when Daniel was a child. As a prophet during the Babylonian exile, Daniel predicted the fall of the Babylonian Empire and the rise of the Medes and the Persians, which he witnessed within his lifetime. But most important, Daniel revealed a prophetic list of all the world empires that would rise and fall before the second coming of Christ. Daniel's prophecy offers a complete outline of the history of world empires written before these empires came into existence. This outline is so important to the understanding of the future of the nations that Jesus related His predictions to events in Daniel's prophecy.

Daniel's outline of world empires included the Babylonian Empire, the empire of the Medes and Persians, the empire of Greece, and the Roman Empire. Daniel also anticipated a final stage to the Roman Empire that is still future. It is this final world empire and its world dictator that will push the world toward Armageddon. In that final world war the history of empires built by conquest will be ended forever by the second coming of Christ. Empires and governments created by men will be replaced by a direct rule of God on earth — a final millennium of peace, righteousness, and prosperity before history ends in a new heaven and a new earth.

The Rise and Decline of the Roman Empire

Daniel named the empires of Babylon, the Medes and Persians, and of Alexander the Great. After the decline and fall of the Greek city states, the next empire to arise was that of Rome, which came into power in the centuries preceding the coming of Christ. This fourth empire was not named by Daniel but was described as "extremely strong; and it had large iron teeth. It devoured and crushed and trampled down the remainder with its feet . . . ." (Dan. 7:7, NASB). The armies of Rome crushed all opposition and extended the iron control of the Caesars over all of southern Europe, western Asia, and North Africa. But Daniel also observed that in the last stage of the empire the iron would become mixed with clay, implying that the fourth kingdom would be vulnerable to sudden destruction (Dan. 2:41-45).

The Future Revival of the Fourth Kingdom

According to Daniel's prophecy, this fourth empire of the Caesars, although seemingly destroyed, is not actually dead and gone. A final form of the last empire is destined to emerge when the leaders of ten nations, originally in the Roman Empire, form a new confederacy of Mediterranean nations (Dan. 7:7). Daniel's vision of this stage of the Roman Empire involved a beast, symbolically representing the Roman

Empire, with the ten horns on the beast representing ten kings yet to arise on the stage of world history. The interpretation of the prophecy was directly given to Daniel, as recorded in Daniel 7:23, 24: "The fourth beast will be a fourth kingdom on the earth . . . . As for the ten horns, out of this kingdom ten kings will arise . . ." (Dan. 7:23, 24, NASB).

The first phase of Rome's rise to power and the empire's division and decline are now ancient history. But the final phase of the fourth kingdom has not yet occurred. This has led students of biblical prophecy to expect a new concentration of power to emerge in the Mediterranean. The fourth world empire will be revived as a final prelude to Armageddon and the second coming of Christ.

The first movements toward the revival of the fourth empire are in today's headlines as the drama of international power and wealth again focuses on the Middle East. Ten strong nations can be expected to emerge in an alliance of political and economic power. That will be the beginning of the future revival of the Roman Empire. This will set the stage for the emergence of the new world dictator.

The Coming Dictator's First Move

A new world dictator will first reveal himself in the role of a peacemaker in the Middle East. This event will take place during the first stage of the revived Roman Empire, the fourth world empire described by Daniel. Symbolically, the new world leader is depicted as "another horn, a little one" who will emerge in the ten-nation Mediterranean Confederacy (Dan. 7:8, NASB). Daniel described his vision of the new world ruler's rise to power in these words: "While I was contemplating the horns, behold, another horn, a little one, came up among them, and three of the first horns were pulled out by the roots before it; and behold, this horn possessed eyes like the eyes of a man, and a mouth uttering great boasts" (Dan. 7:8, NASB).

The ten-nation group which consolidates the power and wealth of the Middle East will soon come under the control of this new leader. Bible expositors have concluded that this emerging ruler will first of all conquer three countries of the original ten and then take control of the entire group of ten nations. The other seven leaders will yield control of the Mediterranean Confederacy to this new strong man of the Middle East.

Daniel identified this man as the one who eventually will become the final world dictator, the Antichrist described in Revelation. He is described as different from the other ten leaders, speaking out against God, persecuting believers in God, and growing in power until he controls the entire world for a period of forty-two months. The interpretation given to Daniel noted that ". . . he will be different from the previous ones and will subdue three kings. And he will speak out against the Most High and wear down the saints of the Highest One, and he will intend to make alterations in times and in law; and they will be given into his hand for a time, times, and half a time" (Dan. 7:24, 25, NASB). The time period of his absolute rule over the earth can be calculated as a year, plus two years, plus half a year — three-and-a-half years (cf. Dan. 12:11; Rev. 13:5).

Understanding Daniel's Seventy Weeks

Daniel's prophecy provides the key to the entire drama of the last days. The new Mediterranean leader will cause a series of world-shattering events, which are described in detail by Daniel and the writer of the book of Revelation. Jesus identified this man as the final military conqueror who would invade Jerusalem and desecrate the temple, "the abomination that causes desolation, spoken of through the prophet Daniel"

(Matt. 24:15, NIV). Jesus' reference to this event added another important piece to the prophetic puzzle.

In Daniel 9:24-27 the prophet recorded a revelation of 70 weeks of years, actually 490 years, in which great events would take place in relation to Jerusalem and the Jewish people. The first two major time segments involved 483 years, or 69 weeks. The word translated "week" actually means "seven" and refers to years not days. This period was described by Daniel in these words: "So you are to know and discern that from the issuing of a decree to restore and rebuild Jerusalem until Messiah the Prince there will be seven [49 years] and sixty-two weeks [434 years]; it will be built again, with plaza and moat, even in times of distress. Then after the sixty-two weeks the Messiah will be cut off and have nothing, and the people of the prince who is to come will destroy the city and the sanctuary. And its end will come with a flood; even to the end there will be war; desolations are determined" (Dan. 9:25, 26, NASB).

The first time segment of Daniel's 70 weeks began with the decree to restore and rebuild Jerusalem, the decree given to Nehemiah in 445 B.C. The first 7 weeks, or 49 years, marked the time needed to rebuild Jerusalem. The second segment of 62 weeks of years (434 years) marked the time which would pass until the Messiah was cut off. Adding these two segments reveals the amazing accuracy of Daniel's prophecy; the 483 years ended just before Jesus was rejected by the nation of Israel and crucified.

But one week of seven years is still left unfulfilled. The "people of the prince who is to come," referring to the Roman army of that day, did come and destroy the city and the temple, as predicted, in A.D. 70. But the future prince, the ruler of the revived Roman Empire, will fulfill his prophetic destiny during the last seven years yet to come.

140

The new Mediterranean leader will be "the prince who is to come." He will seize control of the Mediterranean Confederacy of ten nations. With that power consolidated, he will then make his next decisive move. He will negotiate a peace covenant, guaranteeing Israel's security and bringing peace to the Middle East. According to Daniel, this important move will begin the last 7 years of the predicted 490 years. It will be, in fact, the last 7 years of world history before Armageddon and the second coming of Christ.

"And he will make a firm covenant with the many for one week, but in the middle of the week he will put a stop to sacrifice and grain offering; and on the wing of abominations will come one who makes desolate, even until a complete destruction, one that is decreed, is poured out on the one who makes desolate" (Dan. 9:27, NASB).

The Mediterranean leader will move from a tactic of peace to a tactic of crushing power. After making peace with Israel, during the first three-and-a-half years he will rule as dictator of then ten-nation Mediterranean Confederacy, keeping his covenant with Israel. In the middle of the seven years he will break his agreement. As noted in the previous chapter, this very possibly will coincide with Russia's attempt to invade Israel and Russia's mysterious annihilation. With the world balance of power dramatically in his favor and the world dazzled by Russia's defeat, the Antichrist will show his true colors. He will declare himself world dictator and move to crush all opposition.

Forty-Two Months of Horror

The prophets throughout the Bible have described the last years before the second coming of Christ as a time of great trouble. This is the time when the last world dictator will "devour the whole earth and tread it down and crush it" (Dan. 7:23, NASB). This is the time Jesus described as "a great tribulation, such as has not occurred since the beginning of the world until now, nor ever shall. And unless those days had been cut short, no life would have been saved" (Matt. 24:21, 22, NASB). This is the same period and the same ruler described in Revelation 13. Revelation 13:5 tells us he will be given power and authority to reign for forty-two months. Revelation 13:7 reveals, "He was given power to make war against the saints and to conquer them. And he was given authority over every tribe, people, language and nation" (NIV).

Satan's man of destiny will have forty-two months of power as world dictator. In the process of this rule, God will bring His terrible judgments on a wicked, Christ-rejecting world. The prophetic calendar has been announced for centuries. The die is cast. The Middle East will return to the center of the international stage. The leaders of the ten nations will consolidate the power lost by the fall of Rome. The future Mediterranean leader will await the right moment to upset three of these nations and seize control of the Ten-Nation Confederacy. For three-and-a-half years he will masquerade as a prince of peace, the savior of the world. For the next three-and-a-half years he will use satanic wonders and power to declare himself god and ruthlessly crush all opposition. Near the end of that period the nations of the world will field armies to challenge him. Gripped in a dramatic world war, the armies will converge to begin the suicidal battle of Armageddon, as described in the next chapter.

Tools of World Domination Exist Today

It is most significant that in our twentieth century not only does a need for a world government exist but the tools for establishing such a government are now in our hands. For a world government to exist would require rapid communication. Today, the electronic media, especially use of television via satellite, would be a tremendous tool in the hands of a world dictator who needs instant communication with the entire earth. A universal government also requires rapid transportation. Today, men and arms can be transported to any part of the world in comparatively few hours, something that was impossible in any previous generation. The 1973 airlift of war materials by both Russia and the United States to the Middle East illustrates the potential of modern transportation.

Missile warfare could also be a tremendous tool in the hands of any world ruler. Missiles can be fired to any spot in the world in less than thirty minutes. A ruler with nuclear missiles at his disposal could threaten any portion of the world, blackmailing it into submission with the threat of extinction. No previous ruler in the history of the world had such fearful weapons to enforce his rule.

In the field of economics, Scriptures predict that the world ruler will have absolute control of the economy, and no one will buy or sell without his permission (Rev. 13:17). Today, with the advent of modern computers, for the first time in history this would be possible. A world government could keep financial accounts of all the businesses in the world, controlling purchases and sales and compiling an infinite amount of information about every individual.

New computer technology has already been developed for what is called the electronic transfer of funds. This allows a person to buy and sell without using credit cards, checks, or money in the form of currency. Savings and Loan Associations and some retail stores are already using the system in the United States. In a government-controlled economy the electronic transfer of funds approach would allow complete and total control of all transactions. Each person would have an account on a central computer and be assigned an approved number. When he was paid, his employer would credit his pay to his account by an electronic transfer of funds. When a person wanted to buy anything, the store would simply check his number on the computer and then subtract his purchases instantly by using a small computer terminal. If a world ruler chose to use such a system instead of money — no one could buy or sell unless he gave them an approved number.

World Dictatorship Possible Today

In our modern world all forms of representative and democratic governments will continue to be plagued by overwhelming problems. These will tend to undermine efforts at strong and resolute world leadership. As both domestic and international problems increase, the world will look for a new leader to act decisively to end the turmoil precipitated by the Middle East crisis. Both the need and the tools for the control of the world by one strong ruler exist today. The increasing availability of nuclear weapons, the propaganda power of the world media, and the blackmail power of international economic agreements now make it possible for a world dictator to seize control of the world in a way that would have been impossible in any previous generation. In an amazing way, the necessary ingredients for a world government are present for the first time in the history of civilization. The time may not be far away when Scriptures predicting such a government, written long before one was possible, will have their accurate and complete fulfillment.

12

The Day of World Catastrophe

12

The Day of World Catastrophe

Impending Nuclear Disaster

Many signs today point to imminent, worldwide catastrophe. In one generation mankind has moved from apparent self-sufficiency to a feeling of impotence in the face of mounting world problems. Foremost among these problems are the nuclear arsenals possessed by the United States and Russia. Other nations such as Red China and France are now developing their own nuclear capacities, and smaller nations are increasing the pressure to possess nuclear weapons of their own. The nuclear weapons already produced and ready for use would unleash a tremendous destructive force in an intentional or accidental attack and counterattack. Both sides in such a confrontation would lose hundreds of millions of people in a few days, and attendant fallout would sweep the world, taking its toll of all the nations of the earth. Nuclear war alone, without any other form of catastrophe, could begin a tide of destruction beyond human comprehension. Destruction on this massive and formerly incomprehensible scale is clearly predicted for the end time in the book of Revelation and may be the result of nuclear war.

Fantastic nuclear stockpiles have been amassed as a deterrent, presumably, to international coercion and military attack. At the present time, Russia and the United States have the potential to destroy each other many times over. This has created a stalemate as far as nuclear war between the two nations. Neither nation could use nuclear weapons without being destroyed itself. But soon other nations will have limited nuclear capacity, and no solution for the atomic problem has been reached. As the situation grows more complex, the threat of intentional or accidental use of nuclear weapons becomes more likely. Sooner or later some scientific breakthrough could give one nation or the other a superior advantage, or present weapons could fall into the hands of irresponsible fanatics. Under the stress and strain of crises such as exist in the Middle East, confrontation could trigger a suicidal nuclear exchange. While the world as a whole attempts to ignore this gigantic problem, the prospect of another generation living in peace without some nuclear confrontation is unlikely. Sooner or later the destructive power of nuclear weapons may be let loose upon a world which has never fully faced the inevitability of such a disaster.

Increasing Pollution

Nuclear warfare is only one of our great unsolved problems. World-wide pollution is rapidly making the earth an uninhabitable planet. The air and water are being contaminated at such a rate that soon our oceans will no longer be able to support marine life, and air pollution around the world will seriously affect the health and well-being of billions of people. The United States has made heroic efforts to conquer the pollution problem, but it is one of the few nations in the world to take pollution seriously. The energy crisis has already forced ecological compromise, as industry throughout the world makes greater use of coal and lower grades of oil, which take a devastating toll of the environment. It is only a matter of time until the mounting industrial pollution all over the world reaches the point where human life will be seriously affected. Increasing pollution in the world adds a new and frightening dimension to the prophet's description of the sun being darkened, the moon appearing red, or large portions of the seas becoming red like blood, with the resultant destruction of marine life.

Overpopulation and Starvation

Another massive problem is overpopulation. In the United States population has tended to level off in recent years, but in other portions of the world the population explosion continues. The world population continues to increase at the dizzy rate of 93 million people every year. Social planners have warned that unless drastic changes are made in the birth rate the need for food alone could cause a world-wide catastrophe by the year 2000. It is only a matter of time until the number of people in the world exceeds the capacity of world food production and distribution. This will result in the starvation of millions of people.

The world food crisis was greatly accelerated during 1973 and 1974. The Arab oil producers and other oil exporting countries have made a bad situation worse. The high cost of oil has greatly reduced the supply of petroleum based fertilizers and has made them too expensive for most underdeveloped countries. In addition the gasoline needed for irrigation pumps from the Punjab to Japan was more expensive or not available. Droughts, storms, floods and changes in climate ravaged crops in India, Africa, the Soviet Union and parts of China and the United States.

Although United States food production began to increase in 1975, other countries continued to face serious shortages. Only a few, like the Soviet Union, had the financial resources necessary to purchase large amounts of food. Even now, thousands die of starvation every day. The tensions between nations, between those who have food and those who do not, and all the attendant ills of pestilence and disease that plague underdeveloped countries present insuperable problems for the coming generation.

The Threat of Lawlessness

In addition to these problems, for which man has no solution, are the rising statistics of lawlessness and immorality. Nations as well as individuals no longer operate by any mutually accepted sense of justice or morality. The rule of law which has preserved Western civilization is now seriously questioned. Even just laws are often unjustly applied by the courts. Confidence in man's ability to govern himself by law has been undermined by the radical movements in the 1960s and by Watergate in the 1970s.

The last days are compared in Scripture to the time of Noah, when every man did what was right in his own eyes. The threat of lawlessness and anarchy provides the necessary backdrop for the emergence of a world strongman who will promise to bring peace and order to an increasingly troubled world. In almost every area of human life, all signs point to these problems becoming worse instead of better, with no immediate solutions in sight.

The Coming Time of Jacob's Trouble

In such a context Scripture's teachings regarding a final time of great tribulation, a time of unprecedented catastrophe for the entire world, become much easier to understand. Many of these prophecies were given to Israel and concern the "time of Jacob's trouble," but other prophecies reveal that this will not be "Jacob's trouble" alone. The catastrophes of this period will strike the entire earth and affect the destiny of every nation.

In Scripture this time of trouble is related to the end time, when Israel will be regathered to her ancient land. This is most significant in view of the fact that

Israel today is back in her land. The stage is being set for just such a time of trouble. The prophet Jeremiah recorded, "For, behold, days are coming, declares the LORD, when I will restore the fortunes of My people Israel and Judah. The LORD says, I will also bring them back to the land that I gave to their forefathers, and they shall possess it" (Jer. 30:3, NASB). Immediately after this prediction of Israel's return is the description of her time of trouble. Israel is described as a woman experiencing the pains of childbirth. Her day of trouble was described in graphic terms in Jeremiah 30:7: "Alas! for that day is great, there is none like it; and it is the time of Jacob's distress, but he will be saved from it" (NASB). On the basis of this promise, the progression of events as they relate to Israel in the end time will be her regathering, then her time of trouble, and finally her time of deliverance when Christ returns in power to reign.

Daniel's View of the Time of Trouble

The prophet Daniel indicated the same basic order of events in his description of this period. Important sections of Daniel's prophecy have already been explained in previous chapters, and Daniel 11:40-45, containing an important description of events at the end of the period, will be explained in the next chapter. The prophet's concluding description was given in Daniel 12:1. "Now at that time Michael, the great prince who stands guard over the sons of your people, will arise. And there will be a time of distress such as never occurred since there was a nation until that time; and at that time your people, everyone who is found written in the book, will be rescued" (NASB). The verses which immediately follow indicate that this will occur directly before the second coming of Christ.

The World's Time of Great Tribulation

This great time of trouble in the world will be the

period just before the second coming of Christ to the earth. In Daniel 12:11 it is described as extending for 1,290 days, approximately 3½ years. This is in keeping with other prophecies of Daniel, such as Daniel 9:27, where it is described as half of a 7-year period. Revelation 12:6 speaks of the period as 1,260 days or 42 months (cf. Rev. 11:2; 13:5). Brief though it may be, the 3½ years of the great tribulation include the most awful times of catastrophe and destruction that the world has ever known. Many other passages in the Old Testament give additional descriptions, such as Daniel 7:7, 8 and the explanation of it in Daniel 7:19-27. The great time of tribulation, according to Scripture, will be followed by a time of restoration (see Joel 2:1-11, 28-32; Zeph. 1:14-18; Zech. 13:8 — 14:2). The Old Testament as a whole bears a clear witness to this climactic period of trouble that is unprecedented in the entire history of the world. The problems facing our modern world are only the beginning and are setting the stage for this great time of trouble yet ahead.

Christ's Prophecy of the Great Tribulation

In the New Testament the same theme was continued. The Olivet Discourse, which was Christ's answer to the disciples' question concerning the end of the age, describes this same period (Matt. 24, 25). The Lord Jesus spared no words in describing the awfulness of the period. According to Matthew 24:15, it will begin with "the abomination that causes desolation, spoken of through the prophet Daniel" (NIV), that is, the desecration of the temple which Israel will build in the last days and the stopping of the renewed Jewish sacrifices and ceremonies. This will signal the beginning of the awful period.

Christ advised those in Judea at that time to flee to the mountains (Matt. 24:16). He advised them not to return to their houses to take anything, but to flee with utmost speed. It will be a time of great trial and trouble for those with small children. He instructed them to pray that their flight would not be in the winter, when the cold weather would make it more difficult, nor on the Sabbath Day, when Jews normally do not take a journey and their detection would be easier.

The Human Race Almost Destroyed

Christ concluded in Matthew 24:21, 22, "For then there will be a great tribulation, such as has not occurred since the beginning of the world until now, nor ever shall. And unless those days had been cut short, no life would have been saved; but for the sake of the elect those days shall be cut short" (NASB). Here the same terminology is used as in Daniel 12:1, where the period is called "a time of distress such as never occurred since there was a nation until that time." Like Jeremiah and Daniel, Jesus called the period one that is unprecedented — there has never been a time like it in the history of the world, and there never will be another time like it.

Christ described the days of tribulation as being so severe that if they were not "cut short," meaning "cut off" or "terminated," by His second coming, no one would survive the awfulness of the period, and all would perish. It is "for the sake of the elect" that Christ will come to terminate this period after three-and-a-half years and deliver those who have put their trust in Him. By "the elect," He may have meant Israel as the chosen people or both Jews and Gentiles who will come to Christ in salvation in this awful period. They are the chosen ones, chosen for deliverance at the end of the period, and delivered they will be.

The Miraculous Survival of God's People

According to Romans 11:26, "And so all Israel will be saved, as it is written: The deliverer will come from

Zion; he will turn godlessness away from Jacob" (NIV). Israel as a nation will be delivered from her persecutors and, although subject to searching judgment as described in Ezekiel 20:34-38, those who are judged worthy will enter into the blessedness of the millennial kingdom. These survivors who will have turned to God during the tribulation period will become the parents and grandparents of new generations which will populate the earth during Christ's reign on earth. A similar judgment for Gentiles is described in Matthew 25:31-46, where the sheep are ushered into the kingdom, and the goats, representing the unsaved, are put to death. During the period of tribulation, in spite of terrible disasters and persecution, God will miraculously provide for the survival of His people.

Details in the Book of Revelation

The most detailed description of this time of great tribulation is found in the book of Revelation, beginning in chapter 6 and continuing throughout much of the book. In chapter 19 an unparalleled account is given of the terrible catastrophes and destruction of human life that will occur at the end of the great tribulation. As the prophecy was revealed, the overall sequence of events in the period was given in the description of the breaking of seven seals. As recorded in Revelation 4 and 5, John in his vision of heaven saw a scroll, sealed on its edges with seven seals. When the scroll was unrolled, the first seal was broken, then the second, and, in succession, all the seals through the seventh. These introduced events and situations that describe this general period.

Seven Seals of Horror

The first seal portrayed a terrible and satanic world government which will rule during the period of great tribulation. This was described in more detail in Revelation 13. The second seal described war, for the greatest war of all history will occur near the close of the period. The third described famine, the result of both war and the catastrophes that will overtake the earth. The fourth seal described death, for one-fourth of the world's population will be killed by war, hunger, or roving animals. The fifth seal revealed the extensive persecution of the period, for many who put their trust in Christ will seal their testimony with their own blood. The sixth seal described the coming ecological nightmare. Stars will fall from heaven, and there will be disturbances in the seasons and cycles of nature. The sun will become black, and the moon will become as blood through the haze of destruction. Great earthquakes will move across the earth. Revelation 6 summarizes the entire period as a time when "the wrath of the Lamb" will be poured out on the earth — a time of divine judgment (Rev. 6:16, NIV).

Seven Trumpets of Disaster

The seven seals, however, were only the introduction, the general description of the period. When the Apostle John saw the seventh seal opened, out of it came seven trumpets which depicted specific catastrophes, many of which will destroy large portions of the earth. Great disturbances will affect nature and change climates. A third part of the earth will be consumed by fire — either by a direct act of God or by God's use of other means, such as nuclear disaster or a disturbance of the earth's orbit around the sun. A third part of the sea will become as blood, and a third part of sea life will be destroyed. As the trumpets continue to sound, one catastrophe after another will afflict the earth. Demon possession will be common, as described in Revelation 9, and the torment of those afflicted with demons will be like the torment of one bitten by a scorpion.

The sixth trumpet signaled the advance of a great army from the Orient numbering 200 million (Rev.

9:16). This will be a massive movement of men across Asia, probably led by Red China. China has already developed an army and civilian reserve capable of mobilizing 200 million men and women. As an army like this moves toward the Middle East, it will destroy everything in its path. Startling as is its number, so will be the awfulness of its power to kill men. Its part will be to "kill a third of mankind" (Rev. 9:15, NIV). This apparently will be a human slaughter in addition to that mentioned in Revelation 6:8. The book of Revelation makes it very clear that the combined catastrophes of this period will destroy more than half of the world's population.

Seven Last Bowls of Wrath

With the sounding of the seventh trumpet, a new series of catastrophes was revealed, as recorded in Revelation 16. These were the vials or the bowls of the wrath of God. The symbolism is of a full bowl containing a judgment from God which will be poured out on the earth. As these successive bowls are poured out, great catastrophes will afflict the earth. Men's bodies will be covered with sores and afflicted with terrible pain; all life in the sea will die; the rivers and fountains of water will become as blood; unnatural heat will scorch the earth as the heavens are disturbed in their normal course. The sun will eventually be darkened, resulting in increasing darkness, changes in climate, and destruction of plant life.

The disasters poured out from the bowls of wrath are clearly described as the result of God's direct judgment. These judgments can be interpreted as supernatural acts of God, but some could also be interpreted as the consequences of a nuclear war which God allows. Nuclear war would cause tremendous loss of life, disturbance of the earth's orbit, and disruption of nature's cycles. The cumulative effects of nuclear war would cause human suffering from painful radiation burns

156

and the radiation poisoning of water and food. Even a limited nuclear exchange could precipitate a series of earthquakes and geological activity which would continue a chain of destruction throughout the world.

Unusual demon activity will follow the pouring out of the sixth vial, which will dry up the Euphrates River and will further prepare the way for the great army from the East headed for Armageddon. The final judgment, described in the seventh bowl of the wrath of God, will occur during World War III, as the armies of the earth converge on the Middle East for the final, desperate battle of Armageddon. The entire earth will be literally shaken, its great cities will be destroyed, and the contour of the earth will be changed. Islands will disappear; mountains will be leveled.

This will be the final hour of divine judgment on a world which would not let Christ reign over it. This war and this series of plagues will leave a wake of almost unbelievable destruction of human life and property. It will be exactly what Christ predicted — a time of trouble so great that if it were not terminated by His own coming to the earth, no human life would survive. But this time of trouble will only be the background for God's final dealings with men. It will set the stage for the battle of Armageddon and the judgments which will come when Jesus Christ returns.

13

**Armageddon:
The World's Death Struggle**

13

Armageddon:
The World's Death Struggle

Proclamation of a New World Government

Ironically, the Mediterranean leader will begin his world government by proclamation. Using his consolidated position of power in the Middle East, he will promise a new day of peace and prosperity for all who recognize his leadership. His message will be carried by television and radio to the entire world in one day. The brilliant leader of the Middle East will seem to be the answer for a troubled world, a leader who will enforce peace, settle world problems, and bring plenty to the earth.

This man's absolute control of the world politically, economically, and religiously will give him power such as no man has ever had in human history. His brilliance as a leader will be superhuman, for he will be dominated and directed by Satan himself. But during his three-and-a-half-year rule, he will ruthlessly crush all opposition. His true character is indicated by the titles given to him in Scripture. He is described as "the beast" Rev. 13:1-4), and Satan is described in his true character as "the dragon" (Rev. 12:9; 13:4). This period of world history will be the culmination of Satan's dream to be like God, to control the world, to be the object of worship and adoration, and to secure the absolute obedience of all men.

161

The Rush to Judgment

The new world dictator's blasphemy, disregard of God, hatred of the people of God, and murder of countless believers in Christ will bring down the terrible divine judgments described in Revelation 6-18. Catastrophe after catastrophe will follow as the great tribulation unfolds. The world will become increasingly discontent with the leadership of this world dictator who promised to bring them peace and plenty but instead brought the world one massive catastrophe after another. The wonders and miracles he will perform and his demands to be worshiped as God will only lead to the need for more oppression and persecution of dissenters. In spite of the tremendous power placed in the hands of the world dictator, he will be unable to control the situation. Major segments of the world will begin to rebel against him. Eventually, they will converge on the Middle East to fight it out for international power.

The Outbreak of World War III

The prophet Daniel gave a graphic picture of this situation. After introducing the world ruler in Daniel 11:36-39 as an absolute ruler who will claim to be God, the prophet explained that the armies of the world will rise up in rebellion against him. According to Daniel 11:40, "And at the end time the king of the South will collide with him, and the king of the North will storm against him with chariots, with horsemen, and with many ships; and he will enter countries, overflow them, and pass through" (NASB).

Revolt From the South and North

Daniel's prophecy also described a great army from Africa, including not only Egypt but other countries of that continent. This army, probably numbering in the millions, will attack the Middle East from the South. At the same time Russia and the other armies to the North will mobilize another powerful military force to descend on the Holy Land and challenge the world dictator. Although Russia will have had a severe setback three years earlier in the prophetic sequence of events, it apparently will have been able to recoup its losses enough to put another army in the field. The first battles of World War III will be confused as the armies move back and forth with varied success and failures. Apparently, the world ruler will be able to crush some of the first attempts at revolt and gain some preliminary victories, especially in the South, and he will be able to drive back the invasion from Egypt and Africa.

Red China Makes Her Move

But even as the world dictator appears to gain control of the situation, a report will come of the advancing army from the East (Dan. 11:44). To contend with this advancing horde of 200 million (Rev. 9:16), he will be forced to divert a major portion of his military strength to defend himself. Then further tidings will come of another army advancing from the North. At this point the greatest war of all history, involving hundreds of millions of men, will be set in motion, with the Mediterranean as the major battleground. But this will be more than just another world war, for it will play a special role in preparing the world for the next series of prophetic events.

Satan's Battle Plan

According to Revelation 16:13, 14, the armies of the world will be gathered by demon influence. In a sense, this appears to be a strange contradiction. The world ruler, who was established and aided in his control of the entire world by the power of Satan, now will be attacked by armies that have been inspired to come to battle by demons sent forth from Satan.

In Revelation 16:13, 14 John wrote, "Then I saw

three evil spirits that looked like frogs; they came out of the mouth of the dragon, out of the mouth of the beast and out of the mouth of the false prophet. They are spirits of demons performing miraculous signs, and they go out to the kings of the whole world, to gather them for the battle on the great day of God Almighty" (NIV). It is clear that the unclean spirits, symbolized by frogs, represent demons who are Satan's messengers. They proceed out of the mouth of the dragon, the beast, and the false prophet — the evil trinity that is controlling the world.

The Satanic Trinity

Satan's program is always one of substitution. As Christians in their faith have a triune God — the Father, the Son, and the Holy Spirit — so the forces of evil will represent themselves as triune, with the dragon, or Satan, corresponding to God the Father, the beast corresponding to Christ as the King of kings and Lord of lords, and the false prophet corresponding to the Holy Spirit. Just as the task of the Holy Spirit is to cause all men to worship Christ and the Father, so the false prophet will cause men to worship the beast and the dragon. Proceeding from this evil trinity, the demons will go forth, supporting their work with miracles. These satanic miracles will deceive the kings of the earth into joining the world rebellion with the prospect of gaining world power.

The Battle of the Great Day of God

According to Scripture, the nations will actually be gathered "for battle on the great day of God Almighty" (Rev. 16:14, NIV). Why will Satan organize a world war to disrupt his world kingdom? Satan's desperate strategy will be much more important than his control of the world through the world dictator. Although from their point of view they are gathered to fight it out for world power, the armies of the world will actually be assembled by Satan in anticipation of the second coming of Christ. The entire armed might of the world will be assembled in the Middle East, ready to contend with the power of Christ as He returns from heaven. As subsequent events make clear, the movement will be completely futile and hopeless. The armies of the world are by no means equipped to fight the armies of heaven. Still, Satan will assemble the nations for this final hour, and, in fact, the nations will choose to side with Satan and oppose the second coming of Christ. It will be the best that Satan can do. These events will give the nations their choice and allow Satan his desperate bid to oppose Christ's second coming.

Armageddon

The armies of the world will confront each other, gathered "to the place that in Hebrew is called Armageddon" (Rev. 16, 16, NIV). Armageddon is the Hebrew expression formed from the word "arm," meaning mountain, and "Megiddo," referring to a location in northern Palestine. Mount Megiddo is a small mountain located near the Mediterranean Sea, overlooking a valley that stretches out to the east. This broad valley, some 14 miles wide and 20 miles long, apparently will be the focal point for the marshaling of the armies for the final battle of World War III. Although much too small to contain the millions of men that will be involved, it will be the geographic focal point of the final world war and catastrophe popularly referred to as the battle of Armageddon. Actually, the armies will be deployed for several hundred miles in every possible direction — north, south, and east.

As the armies of the world move toward the final battle of Armageddon, the world war which has erupted will have already claimed millions of casualties. Scripture gives only a meager description of the major movements and battles involved. As the armies converge on the valley of Armageddon, the world will be shaken by the last divine judgment of the series, described as the seventh bowl of wrath.

The Apostle John explained the scene in these words: "Then they gathered the kings together to the place that in Hebrew is called Armageddon. The seventh angel poured out his bowl into the air, and out of the temple came a loud voice from the throne, saying, It is done! Then there came flashes of lightning, rumblings, peals of thunder and a severe earthquake. No earthquake like it has ever occurred since man has been on earth, so tremendous was the quake. The great city split into three parts, and the cities of the nations collapsed. God remembered Babylon the Great and gave her the cup filled with the wine of the fury of his wrath. Every island fled away and the mountains could not be found. From the sky huge hailstones of about a hundred pounds each fell upon men. And they cursed God on account of the plague of hail, because the plague was so terrible" (Rev. 16:16-21, NIV).

To Curse God and Die

The armies of the world will be used as Satan's pawns for this final hour of history. The revolt against the world dictator will assemble the nations in the Middle East. The crescendo of wars and destruction will increase until the entire world is on the brink of ruin. Rather than humble themselves, the leaders of the world will be ready to curse God and die. The final showdown between God and the nations will be about to occur. The prophets revealed that all these events will be the preparation for the great climax of history in the second coming of Christ, the establishment of His kingdom on earth, and the judgment of wicked men who would not have Christ reign over them.

14

Christ's Second Coming to Earth

TIME LINE OF BIBLE PROPHECY

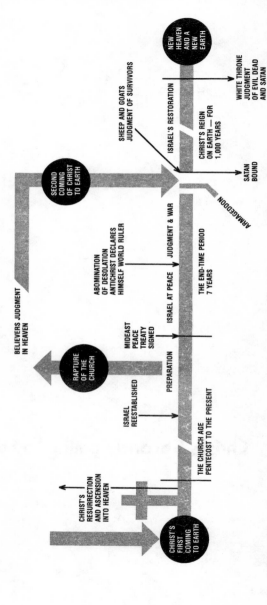

CHRIST'S RESURRECTION AND ASCENSION INTO HEAVEN

CHRIST'S FIRST COMING TO EARTH

THE CHURCH AGE PENTECOST TO THE PRESENT

ISRAEL REESTABLISHED

PREPARATION

RAPTURE OF THE CHURCH

BELIEVERS JUDGMENT IN HEAVEN

MIDEAST PEACE TREATY SIGNED

ISRAEL AT PEACE

ABOMINATION OF DESOLATION ANTICHRIST DECLARES HIMSELF WORLD RULER

JUDGMENT & WAR

THE END-TIME PERIOD 7 YEARS

ARMAGEDDON

SECOND COMING OF CHRIST TO EARTH

SHEEP AND GOATS JUDGMENT OF SURVIVORS

ISRAEL'S RESTORATION

CHRIST'S REIGN ON EARTH — FOR 1,000 YEARS

SATAN BOUND

NEW HEAVEN AND A NEW EARTH

WHITE THRONE JUDGMENT OF EVIL DEAD AND SATAN

14

Christ's Second Coming to Earth

From Manger to Power and Glory

In His first coming to earth, Jesus Christ was born in a stable, relatively unnoticed by the world. Announced by John the Baptist, He came as the servant and Savior described by the Old Testament prophets (see Isa. 42: 1-4; 53: 1-12). He lived in a time of comparative peace, when the Roman Empire controlled Palestine. He was tried and convicted in Jerusalem. At the request of the Jewish leaders, He was nailed to a cross by Roman soldiers. The suffering Messiah who came as a servant was rejected by Israel. After His resurrection and appearance to many, He ascended into heaven from the Mount of Olives near Jerusalem.

The second coming of Jesus Christ to earth will be no quiet manger scene. It will be the most dramatic and shattering event in the entire history of the universe. His coming in power and glory will seize the attention of the entire world. The revived Roman Empire, which will have begun with the world dictator's proclamation forty-two months earlier, will then be torn by rebellion, resulting in a terrible world war. Millions of people will already have been killed in the previous fighting and disasters. Great armies from the north, south, and east will have reached the Middle East, with major contingents gathered in the valley of

Armageddon. Jerusalem, formerly the capital and center of the world dictator's rule, will be under attack, with fighting in the streets. The entire world will then be stopped in awe by the final convulsive shaking of the earth (the last bowl of wrath) and the dazzling appearance of the Son of Man descending from heaven with all His saints.

As Lightning From East to West

In the New Testament, the doctrine of the second coming is one of the major revelations of prophetic truth. It is estimated that one out of every twenty-five verses in the New Testament refers either to the rapture of the church or Christ's second coming to set up His kingdom. Christ Himself described His coming graphically as a glorious event, comparable to lightning coming out of the east and shining to the west (Matt. 24:27). He described His coming as the climax to the tribulation. "Immediately after the distress of those days, the sun will be darkened, and the moon will not give its light; the stars will fall from the sky, and the heavenly bodies will be shaken. At that time the sign of the Son of Man will appear in the sky, and all the nations of the earth will mourn. They will see the Son of Man coming on the clouds of the sky, with power and great glory" (Matt. 24:29, 30, NIV). Many other references to this event are found in the Gospels (Matt. 19:28; 23:39; Mark 13:24-37; Luke 12:35-48; 17:22-37; 18:8; 21:25-28).

Even on the occasion of Christ's departure from earth, the angels reaffirmed the doctrine of the second coming, addressing the disciples, "Men of Galilee, they said, why do you stand here looking into the sky? This same Jesus, who has been taken from you into heaven, will come back in the same way you have seen him go into heaven" (Acts 1:11, NIV). There are many other references to this event throughout the remainder of the New Testament (see Acts 15:16-18; Rom. 11:25-

27; 1 Cor. 11:26; 2 Thess. 1:6-10; 2 Pet. 3:3, 4; Jude 14, 15).

The Whole World Awestruck

The most detailed description of the second coming is found in the book of Revelation itself, which was named for the fact that Christ will reveal Himself to an awestruck world at His second coming. Revelation 1:7 states: "Look, he is coming with the clouds, and every eye will see him, even those who pierced him; and all the peoples of the earth will mourn because of him. So shall it be! Amen" (NIV).

The most graphic picture of all, however, is found in Revelation 19:11-21. As a climax to the great tribulation and the battle of Armageddon, the heaven will break open with the glory of God, fulfilling the prediction of Christ that His coming will be like the lightning shining from east to the west. As revealed in other Scriptures, the heavens will have been darkened by earlier judgments from God, but at that moment an unearthly, brilliant light will spread itself across the heavens, startling the entire earth.

The Rider on the White Horse

As contending armies pause in their conflict with each other, the heavens will open, and Christ will begin the majestic procession from heaven to earth. The words of this prophecy are the most dramatic to be found anywhere in the literature of the world. "I saw heaven standing open and there before me was a white horse, whose rider is called Faithful and True. With justice he judges and makes war. His eyes are like blazing fire, and on his head are many crowns. He has a name written on him that no one but he himself knows. He is dressed in a robe dipped in blood, and his name is the Word of God. The armies of heaven were following him, riding on white horses and dressed in fine linen, white and clean. Out of his mouth comes a sharp sword with which to strike down the nations. He will rule them with a rod of iron. He treads the winepress of the fury of the wrath of God Almighty. On his robe and on his thigh he has this name written: KING OF KINGS AND LORD OF LORDS" (Rev. 19:11-16, NIV). This spectacle will certainly be an awesome one — millions of men and angels reflecting the glory of God and led by Christ astride a white horse, a symbol of a conqueror.

World War III Ended

The world will first cringe in fear before this spectacle, and then the armies on both sides of the great world war will forget their differences and attempt to unite and fight the armies from heaven. This rebellion, inspired and organized by Satan, will be a futile, suicidal effort to resist God. Revelation 19:15 describes the judgment that follows. The sharp sword which goes out of the mouth of Christ is apparently a symbolic revelation of the command that will be issued. All the wicked hosts will be destroyed. The verses which follow reveal that the birds of the air will be invited to feast upon the flesh of the kings, mighty men, and horses slain in this great and catastrophic judgment. The beast, that is, the world ruler, and the false prophet associated with him will be captured alive. According to Revelation 19:20, these two will be cast directly into the lake of fire.

The prophet Zechariah gave additional insight into the events related to the battle of Armageddon. On the very day of the second coming of Christ there will be house-to-house fighting in Jerusalem. The prophet also described the dramatic geological upheaval that will occur when Christ finally sets foot on the Mount of Olives. "For I will gather all the nations against Jerusalem to battle, and the city will be captured, the houses plundered, the women ravished, and half of the city exiled, but the rest of the people will not be cut off from the city. Then the LORD will go forth and fight against those nations, as when He fights on a day of battle. And in that day His feet will stand on the Mount of Olives, which is in front of Jerusalem on the east; and the Mount of Olives will be split in its middle from east to west by a very large valley, so that half of the mountain will move toward the north and the other half toward the south" (Zech. 14:2-4, NASB).

The second coming of Christ will bring an abrupt halt to the final world war and the wave of destruction which will have almost destroyed the earth. His coming will also end the times of the Gentiles, the period during which Gentile nations determined the destiny of Jerusalem and oppressed the people of Israel. A complete understanding of the characteristics and background of Christ's second coming is crucial for every student of prophecy. The descriptions given in Scripture allow a sharp contrast to be made between Christ's second coming to earth and His earlier rapture of the church.

A Personal Return

The second coming of Christ will be a personal return. The same Jesus Christ who was born of a virgin, who died on the cross, who rose again, and who ascended into heaven will soon come back bodily to the earthly sphere to exert His power and sovereignty in the world. It is quite clear that it will be a bodily return, not merely the spiritual presence of Christ. Zechariah 14:4 refers to the fact that "in that day His feet will stand on the Mount of Olives, which is in front of Jerusalem on the east . . . ." (NASB). As the feet of Christ left the Mount of Olives in His ascension, so they will return, but this time in a demonstration of His omnipotent power.

173

A Visible and Glorious Return

The return of Christ will be a visible and glorious return. According to Scripture, everyone will see Him (Matt. 24:30; Rev. 1:7). The second coming of Christ will not take place in a moment, as is true of the rapture of the church. Christ's return will be a spectacular and majestic procession from heaven to earth which will take many hours. During this period, the movement of the procession and the earth's continued rotation will permit the entire world to witness the event. The ultimate destination of the procession will be the center of the Middle East, leading to the destruction of the armies assembled for the battle of Armageddon and Christ's final descent to Mount Zion.

A Return With the Heavenly Host

According to Scripture, a huge body of heavenly hosts, described as the armies of heaven, will accompany Christ in His second coming (Jude 14; Rev. 19:14). Believers who were raptured at the beginning of the tribulation period, who are in heaven with the Lord as promised in John 14:1-4, will return to the earth as part of this vast company. Angels also will join Christ in this great procession from heaven to earth (Matt. 25:31).

A Return Which Shakes the Earth

After the armies of the world are already gathered for battle at Armageddon, the final bowl of wrath, described in Revelation 16:17-21, will be poured out. At that time, immediately preceding the second coming of Christ, will come flashes of lightning, thunder, rumblings, and a severe earthquake. Cities will literally collapse, islands sink, and mountains disappear. Huge hailstones, each weighing a hundred pounds, will fall from heaven. Then, after the procession from heaven and the destruction of the opposing armies of the world, Christ's feet will touch the Mount of Olives

174

outside Jerusalem. At that moment the mountain will be divided, and where the Mount of Olives stands today, a great valley will stretch out into the Jordan Valley below. This is only one of the tremendous changes which will take place in the topography of the Holy Land as a preliminary to Christ's reign. Christ's second coming will drastically change the geography of the earth.

A Return to Judge the Nations

In showing the majestic sovereignty of God in human history, the second psalm gave this description of the world situation at the second coming of Christ: "The kings of the earth take their stand, and the rulers take counsel together against the LORD and against His Anointed: Let us tear their fetters apart, and cast away their cords from us!" (Ps. 2:2, 3, NASB). The rebellion of the nations will draw a derisive laugh from the Lord. "He who sits in the heavens laughs, the Lord scoffs at them. Then He will speak to them in His anger and terrify them in His fury" (Ps. 2:4, 5, NASB). Prophets throughout the Old Testament have pointed to a time when God will speak forth in righteous judgment on the earth. His final purpose is revealed in Psalm 2:6. "But as for Me, I have installed my King upon Zion, My holy mountain" (NASB).

In Christ's second coming, He will claim possession of the nations as His inheritance. His judgment will be absolute, for as the psalmist predicted, "Thou shalt break them with a rod of iron, Thou shalt shatter them like earthenware" (Ps. 2:9, NASB). In light of this predicted judgment and the return of Christ, those who heard were instructed to serve the Lord and worship Him (Ps. 2:10-12). God promised to send His Son at the appointed time to reign over the earth and to begin His absolute rule of peace and justice. And so it will be that the rulers and their armies who resist Christ's return will be killed in a mass carnage. The rebels who are killed in the battle of Armageddon will remain in their common grave until the final resurrection and judgment at the end of the thousand-year reign of Christ (Rev. 19:17-21).

A Return to Rule From Zion

The purpose of Christ's return to the Mount of Olives will be to establish Jerusalem as the capital of His new world kingdom. The law will once more go forth from Zion (Isa. 2:3). Christ's return will save Jerusalem and the nation of Israel from complete annihilation. This direct intervention of God in saving Israel and returning Jews from all nations was predicted as early as the promise recorded in Deuteronomy 30:3: "Then the LORD your God will restore you from captivity, and have compassion on you, and will gather you again from all the peoples where the LORD your God has scattered you."

A multitude of Old Testament passages have developed the same theme of a final kingdom of justice, righteousness, and peace to be established in Jerusalem. In Psalm 24 the gates of Jerusalem are urged to open and welcome the coming King. Psalm 72, in the form of a prayer, describes the reign of the coming King and the blessing and peace that will follow. In this final kingdom, all kings will bow down before Him and the nations will serve Him (Ps. 72:11). Psalm 96 declares that His coming will be for judgment. "For He is coming to judge the earth. He will judge the world in righteousness, and the peoples in His faithfulness" (Ps. 96:13, NASB). Psalm 110 predicts that at the coming of Christ His enemies will be made His footstool, and He will reign in the midst of His enemies. It will be a time when Christ "will judge among the nations, He will fill them with corpses, He will shatter the chief men over a broad country" (Ps. 110:6, NASB).

Many references in the major and minor prophets

can be added to the testimony of the psalms on this subject. According to Isaiah 9:6, the child to be born will be the mighty God and the Prince of Peace; when He comes, "There will be no end to the increase of His government or of peace, on the throne of David and over his kingdom, to establish it and to uphold it with justice and righteousness from then on and forevermore. The zeal of the LORD of hosts will accomplish this" (Isa. 9:7, NASB). The entire eleventh chapter of Isaiah describes the coming of Christ and His judgment and righteous rule of the earth. Many other passages give a glowing description of the glorious kingdom of Christ on earth.

The second coming itself is pictured in Daniel 7:13, 14. "I kept looking in the night visions, and behold, with the clouds of heaven one like a Son of Man was coming, and He came up to the Ancient of Days and was presented before Him. And to Him was given dominion, glory and a kingdom, that all the peoples, nations, and men of every language might serve him. His dominion is an everlasting dominion which will not pass away; and His kingdom is one which will not be destroyed" (NASB).

In a similar way, Daniel 2:44 predicts that at the coming of Christ His reign will be established on the earth. "And in the days of those kings the God of heaven will set up a kingdom which will never be destroyed, and that kingdom will not be left for another people; it will crush and put an end to all these kingdoms, but it will itself endure forever" (NASB).

Living Jews Regathered and Judged

The annihilation of the armies that resist Christ's return will be God's judgment on the nations. After the final carnage of the battle of Armageddon, the surviving people of the world will be judged one by one. All living Jews, the surviving nation of Israel, will be gathered from their hiding places in Palestine and throughout the world (Ezek. 39:28). Each one will face God as his judge, and none can escape this judgment. The rebels who have not accepted Christ as their Messiah prior to this second coming will be put to death (Ezek. 20:38). The rest, believing Jews who have survived the persecution of the tribulation period, will be allowed to enter the Promised Land as the first citizens of Christ's new kingdom on earth. Their hour of persecution will be finished forever, and they will receive all the blessings that have been promised to the children of Israel since the time of Abraham (see Jer. 31:31-34; Rom. 11:26, 27).

Other Survivors Face Judgment

The judgment of the non-Jews who survive the tribulation period was described in Matthew 25:31-46. According to this Scripture, Christ will gather all the Gentile population of the earth to appear before His newly established throne in Jerusalem. In this judgment, individuals will fall into one of two classes — the sheep and the goats. Christ as King will invite the sheep to enter into His kingdom. They are the ones who will have aided the Jews during their intense persecution. They will have visited them in prison, clothed them when naked, fed them when hungry, and hidden them from their tormentors.

The greatest demonstration of true faith in God during this period of intense anti-Semitism will be action taken to help the suffering Jews. While these works in themselves will not be the basis of salvation even in this period, to befriend the Jew during this time will be an unmistakable evidence of true faith in Christ and an understanding of the Scriptures. Accordingly, these believers will be known by their works in a difficult time, but, like the believers of all ages, they will be saved solely by faith in God through the gracious death and redemption of Jesus Christ.

By contrast, the unbelievers will be exposed by a

revelation of their selfish cruelty toward the Jews during the tribulation. These individuals, described as goats, will be cast into everlasting fire — signifying that they will be punished by being put to death. Just as unbelieving Jews will be judged, so, too, unbelieving Gentiles will be judged.

The remaining population of the earth, still in their mortal bodies, will then enter the millennial kingdom of Christ's reign on earth. The kingdom will begin with all unbelievers removed from the earth. As young children mature and new babies are born during the coming thousand-year reign, each of them will also have his moment to believe or rebel. But at the beginning of the period, the rebels will have been completely purged, and the world will enter a new era of peace to rebuild and replenish the earth.

From Disaster to Utopia

A utopian world will follow the colossal failure of man's attempt to control human history. Three judgments will have purged the world of all who have not believed in Jesus Christ. The armies of the world will have been destroyed on the battlefields of the Middle East. Unbelieving Jews will have been judged and killed. In the judgment of the sheep and goats, unbelieving non-Jews will also have been purged from the earth. The entire adult population of the earth which remains will have experienced regeneration through faith in Christ.

Believers of the past will also have a part in Christ's new reign of righteousness. The individuals who placed their faith in Christ during the centuries of the church age will be taken from the earth just before the tribulation period begins. As explained in the next chapter, at the moment of the rapture, the dead in Christ will be raised, and the living will be taken from the earth. In new, resurrected bodies, they will join Christ in the heavens to be saved from the time of judgment on the earth. Believers who died during Old Testament times or during the time of tribulation will be raised and given new, resurrected bodies at the time of Christ's second coming to earth to establish His kingdom. This entire company of saints from the past will be present to observe and help administer the new kingdom on earth.

In a dramatic turn of events, the King of kings and Lord of lords will have seized direct control of human history. A world formerly dominated by Satan and evil men will now be ruled in righteousness and equity. A world torn with war and disaster will enjoy a thousand years of peace. The surviving population of the earth will enter a golden age, when man's intelligence and best energies will be used to rebuild the world and live for the glory of God. Satan will be bound; evil will be directly and swiftly judged by God Himself. Nature will be released from bondage in this balanced and peaceful world where the lion will lie down with the calf.

The Question for This Generation

The present crisis in the Middle East is the beginning of a series of events which will lead inevitably to the second coming of Christ. The first important series of prophetic events in our day came when Israel returned to her land and proved continually to the world that it would not be driven into the sea. The second series of prophetic events was begun when Europe and the world suddenly realized their crippling dependence on Arab oil. Traditional power alignments have now been shaken, and a new confederation of Mediterranean nations seems imminent. As these events continue, the countdown to Armageddon could begin at any time. Our generation may well witness the stirring events described in the dramatic prophecies of the Old and New Testaments. A pointed question is pertinent to each of us in this generation. Are we ready for the coming of the Lord?

Summary Chart of Unfulfilled Prophecy

1. Rapture of the church (1 Cor. 15:51-58; 1 Thess. 4:13-18)

2. Revival of Roman Empire: Ten-Nation Confederacy (Dan. 7:7, 24; Rev. 13:1; 17:3, 12, 13)

3. Rise of Middle East Dictator (Dan. 7:8; Rev. 13:1-8)

4. Peace treaty with Israel: seven years before second coming of Christ to establish kingdom on earth (Dan. 9:27)

5. World church established (Rev. 17:1-15)

6. Russia attacks Israel: about four years before second coming of Christ (Ezek. 38, 39)

7. Peace treaty with Israel broken: world government, world economic system, world atheistic religion begins, three-and-a-half years before second coming of Christ (Dan. 7:23; Rev. 13:5-8, 15-17; 17:16, 17)

8. Martyrdom of many Christians and Jews (Rev. 7:9-17; 13:15)

9. Catastrophic divine judgments poured out on the earth (Rev. 6-18)

10. World war breaks out in Middle East: battle of Armageddon (Dan. 11:40-45; Rev. 9:13-21; 16:12-16)

11. Second coming of Christ (Matt. 24:27-31; Rev. 19:11-21)

12. Judgment of the wicked (Ezek. 20:33-38; Matt. 25:31-46; Jude 14, 15; Rev. 19:15-21; 20:1-4)

13. Satan bound (Rev. 20:1-3)

14. Resurrection of the saints (Rev. 20:4)

15. Millennial kingdom begins (Rev. 20:5, 6)

16. Rebellion at end of millennium (Rev. 20:7-10)

17. Resurrection and judgment of the wicked: Great White Throne Judgment (Rev. 20:11-15)

18. Eternity begins: new heaven, new earth, new Jerusalem (Rev. 21, 22)

15

A Promise to Remember

15

A Promise to Remember

A Heartwarming Promise for Troubled Days

While many dramatic prophecies await fulfillment at the end of the age, none is more important or heartwarming than a simple promise Christ made to His disciples. On the night before His crucifixion as He gathered with the disciples in the upper room, Christ warned them that He was going to leave and that they could not follow Him. One of their number was going to betray Him; another would deny Him. The disciples were deeply troubled. In this tense atmosphere, Christ spoke the memorable words recorded in John 14:1-3. "Do not let your hearts be troubled. Trust in God; trust also in me. There are many rooms in my Father's house; otherwise, I would have told you. I am going there to prepare a place for you. And if I go and prepare a place for you, I will come back and take you to be with me that you also may be where I am" (NIV).

The New Promise Different From
the Second Coming

This coming, often referred to as the rapture of the church, was dramatically different from events Christ had announced in the Olivet Discourse as recorded in Matthew 24. There He had predicted many events preceding His return to the earth. His return to earth would obviously come after the time of great tribulation, beginning when the Jewish temple was desecrated and including a period of intense persecution for the people of Israel. This time of trouble would be dramatically climaxed by the glorious appearing of Christ to set up His kingdom. But the new promise Jesus made to the disciples in the upper room involved an entirely separate event — the coming of Christ to meet believers in the air, the rapture of the church.

The promise of the rapture added a new dimension to the disciples' expectations. Jesus was returning to heaven to prepare a place for them in the Father's house. He would return to receive them to Himself, taking them where He was in the Father's house. This was a promise of removal from earth and entrance into heaven. He mentioned no signs preceding and no tremendous prophecies that needed to be fulfilled first. It was to be an imminent hope, something that would be the expectation of each day after He left them.

Some interpreters of the Bible have attempted to combine this coming of Christ for believers with His second coming to earth with the heavenly hosts from heaven to set up His kingdom (Rev. 19). But many other interpreters have found in careful study of this and other passages that Christ's coming for His own was predicted as an entirely separate event which would precede rather than follow other major end-time events.

Believers today can cling to the same hope that Christ promised His troubled disciples. It would be difficult not to be troubled if Scripture did not clearly show that inevitable martyrdom in the great tribulation would not be the next expectation for believers in this age. True Christians today need not fear the catastrophic days about to overcome the world. Instead, they have the imminent hope of Christ's return and their being joined to the Lord to enjoy His presence forever.

Years after the first announcement of this promise by Christ, the Apostle Paul by divine revelation gave the Thessalonian church additional details on this precious hope (1 Thess. 4:13-18). He described this coming of Christ for His own as a truth that God wanted the Thessalonians to know and understand.

According to Acts 17:1-9, Paul, Silas, and Timothy preached the gospel in Thessalonica for three Sabbath days, with the result that "a great multitude" believed. But unbelieving Jews opposed Paul's ministry and threatened his life, so Paul and his companions left secretly by night to avoid being killed. Sometime later Paul sent Timothy back to see how the Thessalonians were getting along. Timothy's report to Paul gave a glowing testimony of how the Thessalonian Christians, in spite of persecution and opposition, had given a faithful testimony for Christ. Everyone in the region was talking about this Jesus of Nazareth who died and rose again to be the Savior of those who trusted in Him.

But Timothy also brought back theological questions which he could not answer, and one of them apparently related to the subject of the Lord's return. The Thessalonians believed that their loved ones who had died would be raised, but would they be raised before or after the coming of Christ for those still living? Their question reveals that Paul had already taught them about the coming of the Lord for them. For these believers the rapture was an imminent hope, an event that would occur before the tremendous end-time events prophesied in the Old Testament. They simply wanted to know if their dead loved ones would join them.

The Rapture to Include Christians Who Have Died

In reply Paul assured the Thessalonians that God did not want them to be ignorant about this truth or to sorrow like unbelievers, who had no hope. He went on to assure them of the certainty of their hope in I Thessalonians 4:14: "We believe that Jesus died and rose again and so we believe that God will bring with Jesus those who sleep in him" (NIV). This hope of Christ's return for the living and the dead was just as certain as were the facts of the death and resurrection of Christ, in which they had put their trust.

But what did Paul mean when he said that "God will bring with Jesus those who sleep in him"? Paul was here assuming the great fact expounded in 2 Corinthians 5 that when a Christian dies his soul goes immediately to heaven, for to be "away from the body" is to be "at home with the Lord" (2 Cor. 5:8, NIV). The solution given was direct and easy to understand. When Christ comes from heaven to meet believers, He will bring these souls with Him. Their bodies will be raised from the grave at the rapture, and their souls will reenter their resurrection bodies.

Understanding the Order of Events

Paul then described the event: "According to the Lord's own word, we tell you that we who are still alive, who are left till the coming of the Lord, will certainly not precede those who have fallen asleep. For the Lord Himself will come down from heaven, with a loud command, with the voice of the archangel and with the trumpet call of God, and the dead in Christ will rise first. After that, we who are still alive and are left will be caught up with them in the clouds to meet the Lord in the air. And so we will be with the Lord forever" (1 Thess. 4:15-17, NIV).

This new truth, not revealed in the Old Testament, was given to Paul by direct revelation, "by the word of the Lord" (1 Thess. 4:15, AV). Those who are living at the time of the Lord's return will not go before those who have died. Instead, Paul described the scene as the Lord descending from heaven with a shout, literally, a shout of command. Christ will order the resurrection of the dead and the translation of the living.

Christ's Authority to Order Resurrection

This promise is in keeping with what Christ announced to His disciples while on earth when He said, "Do not be amazed at this, for a time is coming when all who are in their graves will hear his voice and come out — those who have done good will rise to live, and those who have done evil will rise to be condemned" (John 5:28, 29, NIV).

Although the rapture will be a partial fulfillment, since it involves only the resurrection of the Christians who have died in this age, it will be the Lord speaking with authority when He gives the shout of command. This will be accompanied by the voice of Michael the archangel (1 Thess. 4:16; cf. Jude 9) and signaled by "the trumpet call of God," or "the last trumpet" (1 Cor. 15:52, NIV). It will be the final call for Christians. They will have heard the call of the gospel and responded in faith. They will have heard the call to service and yielded their hearts and lives to the Lord. Now this will be the last call, the end of their earthly pilgrimages and the beginning of eternity in the presence of the Lord.

A Reunion in the Air

The event of Christ's coming for His church was described dramatically: "the dead in Christ will rise first" (1 Thess. 4:16, NIV). Immediately following, Christians who "are still alive and are left will be caught up with them in the clouds to meet the Lord in the air" (1 Thess. 4:17, NIV). What a reunion that will be! Believers in Christ will see for the first time the One, whom having not seen, they have loved (1 Pet. 1:8). What a reunion it will be for Christians who have been separated by death! Families will be

189

united and friends will see each other again. Their joy at being in each other's presence will be increased by the fact that they will never be separated again. They shall be forever with the Lord.

The Comfort of the Rapture

In the light of this marvelous promise, Paul told the Thessalonians, "Therefore encourage each other with these words" (1 Thess. 4:18, NIV). It is obvious that the comfort offered by Paul had the prospect of immediate fulfillment, that the time of their separation from their loved ones could be short. Such a comfort would be impossible if they must first pass through the traumatic events of the great tribulation, which few would survive. They had the comforting prospect of Christ's coming for them before the world would be engulfed in the maelstrom of the great tribulation. This indeed was a comforting hope.

For the Living — Rapture and a New Body

Some years later the Apostle Paul added further light on this important subject in writing to the Corinthian church. After fourteen chapters of correcting them in various doctrinal and moral matters, in chapter 15 he set forth the great doctrines of the faith. He proclaimed again the gospel that Christ had died for their sins. He supported the great truth that Christ rose from the dead. The resurrection of Christ, he reasoned, is the basis for our hope for the resurrection of our own bodies from the dead. Our present bodies, which are destined for the grave, must necessarily be renewed and made over into heavenly bodies suited for spiritual experience and service in the presence of the Lord. Accordingly, in God's plan the normal course is birth, life, death, and resurrection for those who put their trust in Him.

At the close of the chapter, however, the Apostle declared a mystery, a truth hidden in the Old Testament but revealed in the New Testament (cf. Col. 1:26). The new revelation was that there would be a grand exception to the normal rule of life, death, and resurrection. He declared, "Listen, I tell you a mystery: We shall not all sleep, but we shall all be changed — in a flash, in the twinkling of an eye, at the last trumpet. For the trumpet will sound, the dead will be raised imperishable, and we shall be changed" (1 Cor. 15:51, 52, NIV).

Christians living on the earth at the time of the Lord's return will need to have their bodies transformed just as much as those who have died and been buried. As Paul pointed out, their bodies will be corruptible — subject to decay, age, and all the ills of this life. The bodies of Christians are also mortal, that is, subject to death. As Paul reasoned elsewhere (Eph. 2:1-3 Phil. 3:21), men's bodies are also sinful, not suited for the holy presence of the Lord. A dramatic transformation is necessary, described here as a change to a body that will never grow old and will never die. As Paul expressed it in 1 Corinthians 15:53, "For the perishable must clothe itself with the imperishable, and the mortal with immortality" (NIV). When this is accomplished at the time of the coming of Christ for His own, "Death [will be] swallowed up in victory" (1 Cor. 15:54, AV).

The Rapture Must Occur Before Other End-Time Events

By contrast, of all the many passages that deal with the second coming of Christ to set up His kingdom, while they include the resurrection of martyred dead of the tribulation, none ever mentions any translation of living saints. The reason is that living saints will not be translated when Christ comes to set up His kingdom but rather will be ushered into the millennial reign of Christ in their natural bodies. They will need to lead normal lives, plant crops, build houses, bear children, and repopulate the earth. Such would be impossible if all saints were translated and given heavenly bodies at the time of the introduction of the kingdom. This is one important reason that the rapture must take place earlier in the order of events before end-time troubles overtake the world.

Between the event when the church is caught up to heaven and Christ's return in triumph to set up His kingdom enough time must exist for a new body of believers, both Jews and Gentiles, to come to Christ and be saved. These will be the ones who will populate the millennial earth, while those who have been translated and resurrected earlier at the rapture will join Christ in His reign.

Living in Expectation

As to the Thessalonians, Paul extended to the Corinthian church, in spite of its imperfections, the wonderful hope of Christ's imminent return. As in other references to the rapture, no mention is made of any preceding events. These first century Christians were challenged to be living in expectation of Christ's coming, when the dead in Christ will be raised and living Christians translated. It is this blessed hope (Titus 2:13), this purifying hope (1 John 3:2, 3), and this comforting hope (1 Thess. 4:18) which should thrill the heart of every true believer in Christ. Especially in these momentous days, we should be alert to the challenge of being ready for the coming of the Lord.

16

What Next?

16

What Next?

Watching for the Next Move

As today's headlines report current developments in the Middle East, serious students of prophecy expect the next dramatic and important event to be the rapture of all true believers. Skeptics have been quick to point out that the return of Christ has been the hope of every generation since the time of the apostles. The Apostle Paul clearly hoped that Jesus would return in his lifetime and taught the early church to watch for the Lord's return at any time. Since no specific preceding sign was given to signal the rapture of the church, why has this generation of Christians become so certain that Christ will return in their lifetime? What makes this generation different?

God's Perfect Timing

Scriptures make it plain that, from the divine standpoint, God is never late. If there seems to be a delay in the return of Christ, there must be good and adequate reasons. It is true that over nineteen hundred years have elapsed since the promise of the rapture was first given. The members of the early church in Thessalonica believed that Christ might come in their lifetime, and they were comforted by the expectation of being joined to their loved ones who had already died. But the centuries have passed and generations have come and gone. The world has moved from one crisis to another when it seemed fitting for Christ to return. But the event did not take place.

Why has Christ not already fulfilled His promise to come and receive His own unto Himself? From the standpoint of biblical prophecy, such a delay is not unexpected. When Christ came at His first coming, it was an event which had been anticipated for thousands of years. Centuries of history had prepared the stage for Christ's first coming to earth. The Greek language had developed and was commonly used throughout the Western world. This prepared the way for the New Testament to be written in a precise and widely used language.

The Roman Empire had established comparative peace in the Middle East. Jewish life and worship in Palestine were ready for the Messiah's coming. John the Baptist appeared to announce the Messiah's coming and call the nation to repentance. The Pax Romana had opened international trade and communication, making it possible for the Christians of the first century to spread the gospel message throughout the entire Roman Empire and eventually the world. The prophecies of Christ's first coming were fulfilled in a day carefully prepared by a sovereign God. The prophecies of Christ's return for believers and eventual second coming to earth will be fulfilled in this same way — in God's perfect timing.

Why Has Christ Delayed His Coming?

From the human point of view, it may appear that Christ has delayed His return. But the fulfillment of biblical prophecy in the past has always been preceded by centuries of history which moved with sovereign precision to allow predicted events to occur exactly as promised, complete to the very last detail. But there is also a warm and personal reason why Christ has not returned.

The Scriptures have carefully explained one reason why Christ has not returned. Peter wrote, "The Lord is not slow in keeping his promise, as some understand slowness. He is patient with you, not wanting anyone to perish, but everyone to come to repentance" (2 Pet. 3:9, NIV). The delay of Christ's return to the earth for His own comes from a heart of love. A God of compassion is still waiting for many to hear the Gospel and respond in belief. The delay is for us, so that we can respond in faith and carry this message to others.

The Scriptures indicate that God will not wait forever. Christ Himself used an illustration concerning His coming to establish His kingdom. He compared His coming to the times of Noah. When Noah built the ark in obedience to God's command, more than one hundred years may have elapsed before the flood. Noah attempted to warn his generation, but they scoffed at his message and laughed at his efforts to build the ark. The day came, however, when the ark was finished and the animals went into the ark. Observers could have seen Noah and his family entering the Ark. And then God shut the door (cf. Matt. 24:37-39). Then it was too late! Those outside the ark perished in the flood which followed.

While Christ used this illustration in reference to His second coming, the same analogy applied to His coming for His church at the rapture. God is patient and waits far beyond what human patience endures, but the time eventually comes when God will act.

The Clear Message of Hope

Our generation may well be the last before Christ returns to remove believers from the earth. If so, we have a special burden as well as a special privilege. Those who have confidently received Jesus Christ as Savior have the hope that they may at any moment be called to join Christ in the heavens. For those who have not responded in faith, the same moment will mark the beginning of a series of catastrophic events that will bring God's judgment on the world in which we live.

The burden of our generation is to make God's message of salvation clear in a time of confusion. Living faith brings happiness and meaning into our day-to-day existence. It also provides hope in a world that is rushing to judgment. Each of us must be sure that we have understood and responded to God's message of salvation. Then we must lovingly communicate this urgent message to our families, friends, and everyone we can.

There are many ways to explain the Gospel in making it clear so that people can respond. At the end of this chapter is a statement of the basic message of the Bible from a booklet that has been helpful to thousands of people. The communication of this message is especially important in our generation which may witness the final hour of God's judgment on the world.

Are There Signs of the Lord's Return?

When the rapture of the church was explained, no signs were ever given indicating the specific expectation of this event. Instead, believers in Christ were asked to look for His coming and to anticipate that it could occur at any time. This attitude characterized the early church, which believed that Christ could come any day. This has been the hope of the church from the first century until now. Any suggestion or theory which implies that Christ could not come today is not what has been commonly believed through the centuries. But if there are no signs for the rapture itself, what are the legitimate grounds for believing that the rapture could be especially near for this generation?

The answer is not found in any prophetic events predicted before the rapture but in understanding the events that will follow the rapture. Just as history was prepared for Christ's first coming, in a similar way history is preparing for the events leading up to His second coming. Previous chapters in this book have explained the amazing historical developments that seem to be setting the stage for precisely the predicted events which will occur soon after the rapture. If this is the case, it leads to the inevitable conclusion that the rapture must be excitingly near.

The Three Major Divisions of Prophecy

In the overall study of prophecy three major themes or divisions can be seen in what the Bible says about the future. Prophecies given throughout the Bible generally focus on the future of either the church, the nations, or Israel.

The prophecies that relate to the future of the church explain events and trends which will occur in the true church as the body of believers or in the organized church as an institution. Prophecies about the church as an institution often involve those who are only professing Christians. The Bible has much to say about this professing or organized church and the religious climate of the end of the age. These prophecies reveal the role of the super-church and the false prophet during the tribulation period and the final rule of the Antichrist, who will demand worship as God during the final years of the tribulation.

The rise and fall of nations has been important to prophecies in both the Old and New Testaments. Of special interest today are prophecies that describe the international situation in the final years of history before Armageddon. These prophecies have a remarkable similarity to what can be observed in the world today and, as such, constitute a warning of the approaching coming of the Lord.

Prophecies about Israel, and especially Jerusalem,

provide important reference points for all of prophecy. The most significant prophetic event in the twentieth century has been the restoration of Israel. All the prophecies of the end of the age indicate that at that time the Jews will be back in their land and in precisely the same situation in which they find themselves today.

All areas of prophecy combine in the united testimony that history is preparing our generation for the end of the age. In each area of prophecy a chronological checklist of important prophetic events can be compiled. In each list in regard to the church, the nations, or Israel, the events of history clearly indicate that the world is poised and ready for the rapture of the church and the beginning of the countdown to Armageddon.

A Prophetic Checklist for the Church

When all the predicted events relating to the church are placed in chronological order, it is evident that the world has already been carefully prepared for Christ's return. This checklist includes the major prophetic events in the order of their predicted fulfillment.

1. The rise of world communism marks a new beginning of the politics of atheism.

2. Liberalism undermines the spiritual vitality of the church in Europe and eventually America.

3. The movement toward a super-church begins with the ecumenical movement.

4. Apostasy and open denial of biblical truth is evident in the church.

5. Moral chaos becomes more and more evident because of the complete departure from Christian morality.

6. The sweep of spiritism, the occult, and belief in demons begin to prepare the world for Satan's final hour.

200

7. Jerusalem becomes a center of religious controversy for Arabs and Christians, while Jews of the world plan to make the city an active center for Judaism.

8. True believers disappear from the earth to join Christ in heaven at the rapture of the church.

9. The restraint of evil by the Holy Spirit is ended.

10. The super-church combines major religions as a tool for the false prophet who aids the Antichrist's rise to world power.

11. The Antichrist destroys the super-church and demands worship as a deified world dictator.

12. Believers of this period suffer intense persecution and are martyred by the thousands.

13. Christ returns to earth with Christians who have been in heaven during the tribulation and ends the rule of the nations at the battle of Armageddon.

A Prophetic Checklist for the Nations

The prophetic events related to the nations can also be compiled in a chronological list. This list, too, reveals that the world is dramatically prepared for events which were predicted to occur during the last seven years of the history of civilization.

1. The establishment of the United Nations begins a serious first step toward world government.

2. The rebuilding of Europe after World War II makes possible its future role in a renewal of the Roman Empire.

3. Israel is reestablished as a nation.

4. Russia rises to world power and becomes the ally of the Arab countries.

5. The Common Market and World Bank show a need for some international regulation of the world economy.

6. Red China rises to world power and develops the capacity to field an army of 200 million as predicted in prophecy.

7. The Middle East becomes the most significant trouble spot in the world.

8. The Arab oil blackmail awakens the world to the new concentration of wealth and power in the Mediterranean.

9. France and other Common Market countries move to form strong and lasting ties with Arab oil exporters.

10. United States and Russian influence declines in the Middle East.

11. A world clamor for peace follows the continued disruption caused by the high price of oil, threats of further boycotts, and the inability of Israel and the Arabs to negotiate any lasting peace.

12. Ten nations create a united Mediterranean Confederacy — beginnings of the last stage of the prophetic fourth world empire.

13. In a dramatic power play, a new Mediterranean leader upsets three nations of the Confederacy and takes control of the powerful ten-nation group.

14. The new Mediterranean leader negotiates a "final" peace settlement in the Middle East (broken three-and-a-half years later).

15. The Russian army attempts an invasion of Israel and is miraculously destroyed.

16. The Mediterranean leader proclaims himself world dictator, breaks his peace settlement with Israel, and declares himself to be God.

17. The new world dictator desecrates the temple in Jerusalem.

18. The terrible judgments of the great tribulation are poured out on the nations of the world.

19. Worldwide rebellion threatens the world dic-

tator's rule as armies from throughout the world converge on the Middle East.

20. Christ returns to earth with His armies from heaven.

21. The armies of the world unite to resist Christ's coming and are destroyed in the battle of Armageddon.

22. Christ establishes His millennial reign on earth, ending the times of the Gentiles.

A Prophetic Checklist for Israel

Although Israel's future cannot be separated from the wider sweep of history, prophecies about the Jewish people and the nation have their own distinct order of predicted events. In this prophetic checklist, as in the others, the events already set in motion provide a startling preparation for the final countdown to Armageddon.

1. The intense suffering and persecution of Jews throughout the world lead to pressure for a national home in Palestine.

2. Jews return to Palestine and Israel is reestablished as a nation.

3. The infant nation survives against overwhelming odds.

4. Russia emerges as an important enemy of Israel, but the United States comes to the aid of Israel.

5. Israel's heroic survival and growing strength make it an established nation, recognized throughout the world.

6. Israel's military accomplishments become overshadowed by the Arab's ability to wage a diplomatic war by controlling much of the world's oil reserves.

7. The Arab position is strengthened by their growing wealth and alliances between Europe and key Arab countries.

8. The increasing isolation of the United States and Russia from the Middle East makes it more and more difficult for Israel to negotiate an acceptable peace settlement.

9. After a long struggle, Israel is forced to accept a compromise peace guaranteed by the new leader of the Mediterranean Confederacy of ten nations.

10. The Jewish people celebrate what appears to be a lasting and final peace settlement.

11. During three-and-a-half years of peace, Judaism is revived, and traditional sacrifices and ceremonies are reinstituted in the rebuilt temple in Jerusalem.

12. The Russian army attempts to invade Israel but is mysteriously destroyed.

13. The newly proclaimed world dictator desecrates the temple in Jerusalem and begins a period of intense persecution of Jews.

14. Many Jews recognize the unfolding of prophetic events and declare their faith in Christ as the Messiah of Israel.

15. In the massacre of Jews and Christians who resist the world dictator, some witnesses are divinely preserved to carry the message throughout the world.

16. Christ returns to earth, welcomed by believing Jews as their Messiah and deliverer.

17. Christ's thousand-year reign on earth from the throne of David finally fulfills all the prophetic promises to the nation of Israel.

An Intricately Woven Pattern of Events

The earlier chapters of this book have attempted to present the overall prophetic message about the future as it relates to history and today's headlines. Although prophecies about the church, the nations, and Israel are often presented in separate sections of biblical revelation, they are intricately related to each other in the unfolding of history. Israel obviously had to be reestablished as a nation in Palestine before many of these end-time events could begin. The coming leader of the Mediterranean Confederacy could hardly gain recognition by negotiation of a peace covenant with Israel if Israel did not exist or if there were no Middle East crisis. Without the necessity of Arab oil, European countries could hardly be expected to unite and take an active role in the Middle East crisis. The industrialized nations of Europe now depend on Arab oil in a unique way. This was not true twenty years ago and will probably not be true in another twenty years with alternative sources of energy now being developed.

Many students of prophecy believe that Israel has now completed the first step in her restoration and fight for survival. The next step will require a peace settlement. The Bible predicts that lasting peace will not come in the Middle East until a strong Mediterranean leader of ten nations emerges to establish that peace. The signing of that peace settlement is the second step in Israel's prophetic calendar, and it will mark the beginning of the last seven years before Armageddon. These last seven years will be a time when evil is unrestrained and God's judgment is poured out on the world. The third step in Israel's restoration will begin when her peace treaty is broken. The fourth and final step will begin with the return of Christ and Israel's deliverance.

The prophetic events predicted for the church indicate that these events will not occur until the church, indwelt by the Holy Spirit, is removed from the world. The rise of the Mediterranean leader, the appearance of the false prophet, and the formation of the super-church cannot occur until true believers disappear from the earth. At that point, the Holy Spirit's restraint of evil will be removed from the world, allowing Satan to spin his web of deception and chaos.

For those who believe in Christ, Israel's present situation is tremendously significant. If the next stage in Israel's restoration cannot come until the church is removed, it must mean that the time of Christ's coming for His own is very near. As the Middle East crisis continues to disturb the world, powerful alliances will emerge between European and Arab countries. But the coming world dictator cannot make his decisive move until the church is removed. Only then will Satan have a free hand to manipulate history with the forces of evil. For these reasons, informed Christians are expecting the coming of the Lord as the next important event on the prophetic calendar.

The Final Stage Is Set

The world today is like a stage being set for a great drama. The major actors are already in the wings waiting for their moment in history. The main stage props are already in place. The prophetic play is about to begin.

The Middle East today occupies the attention of the world leaders. The world has now recognized the political and economic power in the hands of those who control the tremendous oil reserves of the area. Old friendships and alliances will be subject to change as European nations seek new alliances and agreements to protect themselves in a changing world situation. The Middle East will continue to be the focal point of international relationships.

All the necessary historical developments have already taken place. The trend toward world government, begun with the United Nations in 1948, is preparing the way for the government of the end time. The world church movement, formalized in 1948, is preparing the way for the super-church that will dominate the religious scene after the true church is raptured. Spiritism, the occult, and belief in demons will continue to spread. Communism, through its atheistic philosophy, is preparing the world for a final form of world religion demanding the worship of a totalitarian dictator.

Israel and the nations of the world have been prepared for the final drama. Most important, Israel is back in the land, organized as a political state, and eager for a role in the end-time events. Today, Israel desperately needs the covenant of peace promised in prophecy. Largely because of the demands of the Palestinians, Israel will not be able to achieve a satisfactory settlement in direct negotiations. Russia is poised to the north of the Holy Land for entry in the end-time conflict. Egypt and other African countries have not abandoned their desire to attack Israel from the south. Red China in the east is now a military power great enough to field an army as large as that described in the book of Revelation. Each nation is prepared to play out its role in the final hours of history.

Our present world is well prepared for the beginning of the prophetic drama that will lead to Armageddon. Since the stage is set for this dramatic climax of the age, it must mean that Christ's coming for His own is very near. If there ever was an hour when men should consider their personal relationship to Jesus Christ, it is today. Many people have found the outline in a booklet published by Dallas Theological Seminary helpful in reviewing their relationship to Jesus Christ and preparing for the future. This booklet is reproduced on the following pages. God is saying to this generation, "Prepare for the coming of the Lord."

how to have a happy and meaningful life

some people feel

that a happy and meaningful life consists of

- a good time
- friends
- lots of money
- doing something worthwhile
- getting to the top

All of these make sense...

but

we all know people who have these things and they are *still empty*.

why?

They started at the wrong place.

Here are four vital facts we need to know.

1 A HAPPY AND MEANINGFUL LIFE BEGINS WITH GOD!

According to the Bible, look what God offers you.

- **LOVE** — Someone who cares about you.
- **SECURITY** — Someone who cares about what happens to you.
- **PEACE** — Someone who cares about your problems.
- **PURPOSE** — Someone who cares about whether your life counts.
- **ETERNAL DESTINY** — Someone who cares about your future.

Jesus said
"God so loved the world, that He gave His only begotten Son, that whoever believes in Him should not perish, but have everlasting life."   John 3:16

Jesus also said
"I have come that they might have life, and that they might have it more abundantly."   John 10:10

Augustine observed
"Thou hast made us for Thyself, O God, and the heart of man is restless until it finds its rest in Thee."

unfortunately...

2 MAN'S SIN HAS SEPARATED HIM FROM GOD

According to the Bible sin is
- Failure to be what God wants us to be
- Failure to do what God wants us to do

Obviously then, all of us have sinned. Consciously or unconsciously we have rebelled against God.

That's not all. God is righteous and cannot allow sin in His presence. God must judge man's sin and reject it.

The Bible says
"All of us, like sheep, have gone astray; we have turned, every one of us, to his own way…"   Isaiah 53:6

"Your sins have been a barrier between you and your God."   Isaiah 59:2

As long as sin separates us from God, you see, we cannot enjoy the happy and meaningful life God wants us to have.

fortunately…

3 GOD LOVES YOU VERY MUCH! HE GAVE JESUS CHRIST, HIS SON, TO TAKE AWAY YOUR SIN.

The Bible says
"God showed His love toward us, in that, while we were yet sinners, Christ died for us."   Romans 5:8

"Christ has once suffered for sins, the just for the unjust, that He might bring us to God."   I Peter 3:18

God is satisfied with what Jesus Christ has done. Now He is completely free to forgive you, and to offer you a happy and meaningful life with Him.

Jesus Christ has removed the barrier of sin through His death.

HOWEVER, YOU HAVE *ONE* RESPONSIBILITY

4 TO ENTER INTO A HAPPY AND MEANINGFUL LIFE, YOU MUST TURN TO GOD BY TRUSTING JESUS CHRIST TO FORGIVE YOUR SINS.

Look at what the Bible tells us
"Jesus said, I am the way, the truth, and the life: no man comes to the Father, but by me."   John 14:6

"Believe on the Lord Jesus Christ and you will be saved."
Acts 16:31

"Being justified [set right with God] by faith, we have peace with God through our Lord Jesus Christ."   Romans 5:1

BUT WHAT IS FAITH?

We exercise faith when

> we *depend* on a doctor
> or *trust* in a lawyer
> or *believe* in a friend

Faith in the Lord Jesus Christ, therefore, is *trusting* Him to forgive your sins and to bring you into a right relationship with God.

let's review...

You know that...
1. Life begins with God.
2. Your sin has separated you from God.
3. Jesus Christ paid the penalty for your sin.

THE ONLY THING GOD ASKS YOU TO DO

4. Trust the Lord Jesus Christ to forgive *your* sins so that you may begin a happy and meaningful life with God

Are you still here? Or have you trusted Jesus Christ completely to forgive your sins?

Wouldn't you like
to trust Jesus Christ
to forgive your sins
right now?

YOU MAY WANT TO USE THIS PRAYER TO EXPRESS YOUR DECISION

"Dear Father, I admit that I am a sinner. I believe that the Lord Jesus Christ died for me. Thank you for forgiving my sins. Let me start a new and meaningful life with You today.   Amen."

Jesus is not a dead Savior. He is *alive*. His power is now available to you. If you have trusted Christ, God has promised to give you His life, and God keeps His promises.   (John 3:15, 16)

Only Jesus Christ can live the Christian life! As a Christian, Christ's Spirit now indwells you. Power in your new life comes as you learn to rely on Christ to live His life through you.
(Galatians 2:20)

That's why you need the Bible. You need to study it daily to become aware of what God wants to do through you.
(I Peter 2:2)

Prayer is another vital part of a happy and meaningful life. Speak to God often about your needs.   (Philippians 4:6,7)

One more thing. You can't be a Christian alone. Meet with others who have trusted the Lord Jesus Christ in a church where the Bible is taught.   (Hebrews 10:25)

From the booklet, **"How to Have a Happy and Meaningful Life."** Single copies free on request from Dallas Theological Seminary, 3909 Swiss Avenue, Dallas, Texas 75204.

"The film, Armageddon, featuring Dr. John F. Walvoord, and dramatizing the chronology and sequence of events leading up to earth's final hour, is available from Space Age Communications Educational, Box 11008, Dallas, TX 75223."